Human. Nature.

FORT WHYTE
RAILWAY
MAP OF
MANITO
LINES IN OPERATION
PROJECTED LINES
Scale of Miles
LAKE WINNIPEG
CANADIAN PACIFIC MAIN LINE
CANADIAN PACIFIC SOUTH WESTERN BRANCH
CANADIAN PACIFIC PEMBINA BRANCH

Human. Nature.

THE FORT WHYTE STORY

JAKE MACDONALD

FortWhyte Alive
HUMAN. NATURE.

1961 McCreary Rd
Winnipeg, MB Canada R3P 2K9
www.fortwhyte.org

Written by Jake MacDonald
Design & Typography by Relish Design Studio Inc.
Printed in Canada by Friesens

Library and Archives Canada Cataloguing in Publication

MacDonald, Jake, 1949-
The Fort Whyte story : human.nature / Jake MacDonald.

Includes index.
ISBN 978-0-9813732-0-1

1. Fort Whyte Foundation--History. 2. FortWhyte Alive (Organization)--History. 3. Fort Whyte Foundation--Biography. 4. FortWhyte Alive (Organization)--Biography. 5. Nature conservation--Manitoba--Winnipeg. I. FortWhyte Alive (Organization) II. Title.

QH77.C3M313 2009 333.7071'071272 C2009-905911-8

To the legions of volunteers and donors,
on whose broad and generous
shoulders Fort Whyte stands

TABLE OF CONTENTS

INTRODUCTION

In a remote valley on the southern slopes of the Pyrenees the Spanish villagers have a proverb:

"A beautiful child has many parents."

This proverb is still chuckled over by the old folk as they sit on their sun-dappled benches around the village square. They have the leisure to ponder such things in a village where our fast-paced world rarely intrudes and time is still reckoned from the day in the Second War when a damaged Lancaster bomber pursued south by German fighters crashed in their mountain valley, and they buried the young RAF crew in a special place in their village cemetery.

Their proverb has as many layers as a good Spanish onion. The outer layer has a little extra bite to it, suggesting with a sly grin that the official records do not always disclose the child's true paternity.

The next layer teases us about our very human inclination to claim parentage of a beautiful child and thereby gain some vicarious credit for its virtue. Welcome aboard the Fort Whyte bandwagon. There is more than enough credit to go around.

But deeper and sweeter in the heart of the onion is the age-old truism that it takes a village to raise a beautiful child. This is where the proverb really hits our mark because our village has done a remarkable job of raising Fort Whyte.

Alan Scarth and Sam Fabro
in Fort Whyte cattails, Fall 2008

The Spanish have no copyright on this proverb. The Italians have a version that was famously enunciated by Mussolini's son-in-law: *"La vittoria trova cento padri e nessuno vuole riconoscere l'insuccesso".*

Sam Fabro from his seat on the front bench overlooking our village square would translate this as: *"Success has a hundred parents, but failure is an orphan"*

Sam Fabro is one of the legitimate parents of Fort Whyte. During the critical first twenty years of Fort Whyte's childhood (which we can truly call the "Sam Fabro Era") Sam chaired the Fort Whyte Management Committee which in some difficult financial times saved our child from becoming an orphan. He made sure our Board of Trustees (and especially its undersigned Chairman) kept their feet on the ground, and that their grand ideas stayed harnessed by the principles of sound management. As a prominent and experienced practitioner of good governance of non-profit organizations, Sam was the steady patriarch of our beautiful child.

When forty long years ago Fort Whyte's pioneers came up with their off-the-wall idea…the delivery of what they called "nature education" to the school children of Manitoba…they were met by some quizzical looks. Environmental education programs were not an easy sell in the Manitoba of the 1960s and even the word "environment" was an unfamiliar one.

The business community, justifiably skeptical of what looked like an unbusinesslike proposition, bought in not because they embraced the concept but because they respected the reputations and the judgment of the first Trustees who were installed in office by the Manitoba Legislature in 1966.

Founding Trustees:

Donald W. Muir	Alan W. Scarth
H. Stewart Perdue	Clarence Tillenius
George T. Richardson	George E. Tritschler
N. Elliot Rodger	

After its early years of growing pains, Fort Whyte graduated from its McGillivray Blvd. quarters and built Carl Nelson's beautiful building in the midst of its newly acquired 220 acres of lake country, triumphally capping the Sam Fabro Era.

The second twenty years of our forty year saga (the "Bill Elliott Era") began when we could at last afford a full-time CEO to replace Sam's crew of hard-working part-timers. After a short apprenticeship Bill took over the chores of Sam's Management Committee which gracefully retired along with the original Trustees. In the ensuing twenty years Bill and his new Board of Trustees have tripled Fort Whyte's land base and tripled its building space without leaving any significant debt on the books.

The story of the combined forty years in the Fort Whyte wilderness is really two stories. It is first and foremost a story of hundreds of good citizens dedicated to Fort Whyte, and this story Jake MacDonald has superbly captured in his book. But it is also a story of good management and good governance, dull things which are often overlooked and which non-profit organizations have not been very good at. A more pedestrian story to be sure, but worth remembering because Sam and Bill's good management created the sturdy fabric into which Jake's colourful human strands are woven.

Veterans of the non-profit battles like Sam Fabro, who has been in more of them than anyone else in our fair province, make no apology for talking about, even harping upon good governance and sound financing, because in the course of our forty years many other non-profit orphans were reduced to begging at the doors of business and government.

Business corporations have the advantage that their shareholders are looking over the shoulders of their boards and management to make sure they are producing profit and dividends. There is a natural and healthy tension between management and the directors on the one hand and on the other the shareholders who hold the executive feet to the fire by deciding how much their shares should sell for. But in non-profit organizations the membership has no money stake, so the members only become interested in governance when they find the front doors closed.

There are no mysteries here. A corporate structure is after all nothing more than a vehicle to take the organization where it needs to go. But it is essential that there be a continuing conversation between the drivers in the front seat and the passengers in the back seat about where we're headed. Without shareholders in the back seat, non-profit organizations like Fort Whyte must create a bi-polar structure with a manager at the wheel and an independent and questioning Board of Trustees in the back seat, the Trustees not questioning every turn of the wheel but having some clear ideas about the destination.

So far, whether it's been Sam or Bill at the wheel, the Fort Whyte governance system has worked, but as we move ahead into the next forty years we should never forget how important it is to have those long-sighted back seat drivers.

And those next forty years look to be even more interesting than the first forty. When we began delivering our message to 5,000 or so school students who came each year to the old McGillivray classroom-on-stilts, the curriculum was called "Nature Education" and it conveyed the revolutionary ideas that all ecosystems in the precious and precarious biosphere of this planet of ours are interconnected and that human beings have the greatest potential in the history of life on the planet to foul those ecosystems up. By the end of our first forty years the 6.5 billion of us human beings have managed to prove that Fort Whyte's warning was soundly based by largely ignoring it and thoroughly fouling our ecosystem nest through our indulgence in fossil fuels.

FortWhyte Alive is now gearing up to deliver a new message for the next forty years, which will include the warning that unless we do something about it, the carbon we are loading into the atmosphere will change our climate and endanger the futures of those young people who in their hundreds of thousands have come trustingly through our doors.

But that is another book for another time. Today let's take a moment to look over our shoulders and deliver some well-earned salutes.

First and foremost we salute our contributors and volunteers, the hands-on parents in our Fort Whyte village.

And we salute the members of the Fort Whyte Legacy Society, who are committing to leave legacies to Fort Whyte's Endowment Fund to better secure our financial base for the next forty years.

And finally and gratefully, we salute the men and women who have toiled in the engine-room of our great organization: Sam Fabro and the Members of his Management Committee, Bill Elliott and his staff, and the Trustees past and present, all of whom have lent their time, their reputations and their wisdom to our cause.

–Alan Scarth

CHAPTER 1

THE FORT THAT NEVER WAS

When people visit FortWhyte Alive for the first time they sometimes look around in a perplexed way and ask a familiar question – "So where's the fort?"

The short answer is, "There is no fort…but there's a good story!"

The name "FortWhyte" originates with William Whyte, a Scotsman who immigrated to Toronto in 1863, landed a job as a brakeman, and began a long and successful career that took him from the bottom of the ladder all the way to the top, as Vice President of the Canadian Pacific Railway.

William Whyte

As the CPR's boss of western development, it was Whyte's job to enforce and protect the railway's national monopoly. Whyte believed that the best way to do this was to concentrate on pushing the rail line out to the Pacific Ocean and thereby ensure that the CPR became the country's first sea-to-sea railway. This put him in a position of conflict with the Manitoba government, that wanted the CPR to help prairie farmers by spending time and resources building branch lines to help them transport their goods to market.

In 1881, Manitoba Premier John Norquay announced that the Province would pursue this goal by building its own railways. The CPR cried foul in Ottawa, and the Conservative government of John A. MacDonald took the side of the big railway. The CPR sought injunctions against the building of a so-called "Northern Pacific and Manitoba Railway" but the track layers kept building, and the two railroads finally converged at a spot close to where the community of Fort Whyte now stands.

Both companies swore to stop the other, and a volunteer police force hustled out to the crossing site to protect the Manitoba crew. They found that the CPR had fenced the crossing site and dumped a locomotive in the path of the Manitobans. The Manitoba crew kept working, under cover of darkness, and the next day the CPR crew tore up their tracks, protected by a squad of 200 CPR men, some armed, who arrived on a special train provided by William Whyte. Every night, as construction proceeded at a stop and go pace, the two armed groups glared at each other over a row of blazing bonfires. A battle was only averted when the Supreme Court ruled that the Manitobans had the right to pursue alternatives to the CPR. So the dispute was settled in a civilized manner. But the local newspapers, with a characteristic zeal for controversy, dubbed the armed camp "Fort Whyte" and the name remains with us today.

The Manitoba of those days was a wide and empty land, with huge swaths of unspoiled prairie where remnant groups of wild buffalo still roamed. Sitting Bull had only recently wiped out the Seventh Cavalry at the Little Big Horn River. And on the Canadian side of the medicine line, Louis Riel and Gabriel Dumont had yet to fight it out with General Middleton and his "Little Black Devils" at the Battle of Batoche.

The Battle of Fort Whyte, 1888

The Royal Winnipeg Rifles wore dark green coats that looked black at a distance. At the Battle of Fish Creek, four years after the standoff at Fort Whyte, a captured Metis combatant asked, "the red coats we know, but who are those little black devils?" The name stuck, and is still proudly used by the regiment, whose motto is *Hosti Acie Nominati*, which roughly translates as "named by the enemy in battle."

Like it or not, European style civilization was coming to Manitoba, and it was coming via the railroad. Manufactured goods in those days mostly originated in Europe, and freight therefore moved by ship across the Atlantic and then by rail across North America. Every axe, saw, leather harness, girdle and teacup had to be loaded onto a train and shipped west, to where the fast-growing agricultural industry was turning the prairie provinces into "the bread basket of the empire." As the main railway distribution centre west of Toronto, Winnipeg was well-positioned to become one of the most important cities in Canada and indeed North America.

Before the days of federal control of the grain trade, Winnipeg was also the centre of the grain industry, and entrepreneurs, capitalists and labourers were flooding into the province to ride the boom. During the last years of the 1800s and the first decade of the 20th century, Winnipeg grew rapidly, with a population that soared from 25,000 to 200,000. As the third largest city in Canada, it also became the centre for banking, education and textiles, and these institutions attracted their own big names. By 1914, Winnipeg had reached its frenzied zenith of development. Along Broadway, the Legislative Building, the Law Courts Building and the Fort Garry Hotel were all under construction. Hundreds of grand apartment buildings and palatial homes were springing up around the city, and Winnipeg's future as "the Chicago of the north" seemed assured.

A look west down Broadway Avenue, Winnipeg

The construction of the Panama Canal, which also began in 1914, eventually made the transportation of freight by steamship less expensive than shipping by railway. And as soon as the Atlantic Ocean became linked with the Pacific by water, Winnipeg's importance began to diminish.

HOMESTEADERS

Grasslands to Farmlands – 'Sod Busting' marked the beginning of the end of the vast native prairie

You can't build a grand city without raw materials. And in this regard Winnipeg was also blessed. Limestone and lumber were available in abundance, as were the ingredients for making cement. Since the recession of the glaciers, 10,000 years ago, floods have regularly spilled out across the Red River Valley, coating the land with beds of clay, which was used to manufacture Portland cement. In 1911, a company called Canada Cement saw an opportunity to cash in on Winnipeg's construction boom and built a cement plant at Fort Whyte. Men with horse-drawn scrapers began stripping off the topsoil from the aspen prairie, then a steam shovel moved in and scooped out the clay, loaded it on small rail cars, (the tracks for which can still be seen), and hauled it back to the plant where it was utilized for making cement. The clay bed petered out about 12-15 feet underground, so the steam shovel moved sideways across the land, creating a series of enormous, wide excavations that quickly filled with spring runoff and rainwater to resemble a chain of prairie lakes.

The famous Winnipeg flood of 1950 may have helped pike, perch and carp find their way into these "lakes" through backed up ditches, but it was surely active stocking that added the walleye and largemouth bass. There are many creeks and rivers in the Winnipeg vicinity, but no natural lakes, so the area became a popular recreational site for those lucky enough to live nearby.

Vicki Daubaras grew up in a family of five in the village of Fort Whyte. When she was a child they heated their house with coal, melted snow for drinking water and fished in the lakes in front of the cement plant. "My dad always caught lots of fish," she says. "He made a scraper by nailing bottle caps onto a board and used it to get the scales off the fish. We'd have them for dinner and they were delicious!"

By most accounts Canada Cement was a good place to work. It relied on its skilled workers, and to encourage long-term employment it provided its workers with everything from low-cost housing to good recreational facilities. Fort Whyte, under the wing of Canada Cement, developed into a special recreational area and a distinct community. It was just a few miles outside of Winnipeg – close enough to provide access but far enough that it felt like a unique town. "It was a big adventure to walk the three miles to Pembina Highway," says Vicki. "We didn't have a car so we'd walk along Clarence to the street car and go to a movie. We had concerts at the company clubhouse and picnics by the lake. I remember one fellow with a weird sense of humour brought a skunk to the company picnic. In the summer we'd get buckets and walk along the trails and pick Saskatoon berries and sell them to our neighbours for twenty-five cents a pail. We saw lots of birds and rabbits but in all those years of walking the woods and fields, I don't recall ever seeing a deer."

Canada Cement was the mother corporation that owned the homes, ploughed the roads and paid the salaries of almost everyone who lived in Fort Whyte. In the morning, a whistle blew to summon the company workers, and at the end of the day the whistle blew to send them home again. When the Great Depression arrived in the 1930s, Winnipeg's building boom came to an end, but Canada Cement soldiered on, and its employees considered themselves lucky to have jobs. Picnics, parties and other social events happened regularly at Fort Whyte, and the company encouraged workers to come up with their own ideas about making the community a better place to live and work.

Wild Ginger

NATURE NOTE

First Nations people taught early European settlers to peel the root of wild ginger and use it as flavouring for foods and beverages. In addition to using the plant to spice up food, some First Nations groups also used wild ginger to treat heart problems, coughs, colds, and earaches.

Canada Cement employee
Archie Carmichael
wets a line in Pit One

Canada Cement dragline

George Nelson

George Nelson lived at Fort Whyte in those days, in a company house along McGillivray Boulevard. In the evening, he would see his grandfather coming home from work coated with cement dust. "It covered his clothes and hair, and as I particularly remember, his bushy eyebrows. Even after washing, the dust seemed to never leave the aged wrinkles on his face."

George Nelson and his chums used to play at "Pit One," the first of the great open-pit clay quarries that the dragline had excavated. They swam, fished and trapped rabbits in the woods around its edges. "I remember the day that my friend Tom Milnes showed up for swim practice at the YMCA, smelling strongly of the skunk glands he had used to disguise the human smell of the traps and snares he had set in the woods around the pits. It caused a bit of a fuss but we didn't want to be city boys. We wanted to be brave trappers and outdoorsmen."

Nelson and his buddies spent many an afternoon wandering the company grounds, camping out in the woods, and sneaking into the pits for an illegal swim now and then. The pits had steep banks, with an overhang cut by the scoop of the dragline. It was dangerous swimming in the pits because of this overhang, and a swimmer who was unskilled or tired might have a nasty surprise waiting for him. Peter Muir was another Fort Whyte kid who also slipped into the lakes for a swim now and then, and recalls that it was important to get as far as possible from the cement plant, beyond the watchful eye of plant employees. "The dangerous place was about or twenty or thirty feet from shore, right at the place where the water was about chest deep on a twelve year old boy," he recalls. "One step more and they plunged way over their heads. Some of those not expecting the change went down and were caught under the overhang and drowned."

NATURE NOTE

The White-tailed Deer, originally a species of Southeast North America, has increased its range north and west over the last century, as human activity has subtly altered the landscape. White-tailed Deer are not open prairie animals, but prefer a mix of fields and wooded shelter. Farmyards, suburbs, parks, and shelterbelts all are human-made changes to the landscape favouring the deer.

Deer

White-tailed deer

Mule deer

Because of the danger, Canada Cement built a fence around the lakes, but kids still persisted in sneaking in to swim and explore. George Nelson undertook overnight camping trips into the woods around the plant, and he remembers them with particular nostalgia. "I remember waking up to an early morning breakfast of thick dark elk sausages with eggs and toast, cooked over an open fire. We used to camp in a patch of woods where the Interpretive Centre now stands, and when I visit there today it's hard to believe what an incredible job has been done in preserving those woods and even improving on the wildlife aspect. My friend Tom died in an accident not long after our camping trip, but I know he would be pleased to see what became of our woods."

After George Nelson grew up, got married, and started a family, he began hunting and fishing, and joined the Winnipeg Game & Fish Association. There he met a man named Don Muir, who happened to work for Canada Cement. Don was a stationary engineer whose job it was to supervise the kiln in the cement plant. George and Don shared an interest in the outdoors, and they sometimes passed the time in the Muir kitchen in Fort Whyte, drinking coffee and talking about their hunting and fishing trips.

In those days, being a nature lover meant that you were probably also a hunter and fisherman, and Muir, a capable outdoorsman, was more enthusiastic about fishing and hunting than most. He organized his fellow employees at Canada Cement into a group called "The Lucky 13 Rod and Gun Club" and they regularly got together for recreational outings and nature conservation projects. In today's world, "hunter" and "Nature enthusiast" are not necessarily synonymous. But in the 1950s, hunting and fishing were the main outlets for men who were keen on Nature. And in any case, Muir and his fellow outdoorsman were not motivated by the belief that Nature must be fenced off and protected from human consumption. They believed in "conservation" as opposed to "preservation." They believed that Nature, in the form of plants and trees and fish and wildlife, was meant to be used by humans, but used respectfully and wisely. With careful management, wild spaces and species would always be there for future generations.

Canada Cement fish derby and wiener roast at Fort Whyte, 1959

In 1957, the Club bought six captive Canada geese with clipped wings and made a safe home for them on the shore of Pit One behind the cement plant. Then they acquired a flock of 50 mallards from Peter Ward of the Delta Waterfowl Research Station. The following year, the Fort Whyte flock was supplemented with 40 geese from the Gabauer family of Morris, Manitoba. In their spare time, the Club members, under Don Muir's supervision, built winter and summer pens for the birds and a building to serve as a feed storage and brooding site. By the early 1960s, Don and his colleagues had a substantial, noisy and very hungry flock of waterfowl on their hands.

Don had such a fatherly attitude towards the birds that he rarely took an evening or a weekend away from them. When he wasn't working, he was over at the waterfowl pond, where he fed the birds, doctored the sick ones, and built sturdy fences and shelters to protect them from predators. His son Peter recalls that when he was very young he would help his dad feed the geese. "Whenever I happen to smell fresh grain it takes me back to those days. I was Dad's retriever. He'd send me into the pens to catch the geese if he needed to get hold of one of them for some reason. They were really big, and they'd give me quite a wrestle. I'd grab them by both wings and bring them to my dad. The pens were only a five minute walk from our house and we'd go over there to feed the birds at least twice a day."

The geese and ducks were beginning to draw the interest of the public, which pleased Don and the other members of the Club, and they encouraged Boy Scouts and Girl Guide groups to come out and see the birds and take a walk in the woods. Several school principals saw the opportunity to incorporate a visit into the school science curriculum, so the members of the Club got together and cut a nature trail. The first school groups were primarily kids in Grade VI and VII, but it soon expanded to other age levels, and the teachers tailored the hikes to suit the interests and attention spans of the children. Very young children of primary school age seemed most intrigued by the live birds. Older kids liked roaming the woods studying animal tracks and learning about animal behaviour. Members of the general public were also enjoying the nature trail, and Don Muir was beginning to realize that their little waterfowl sanctuary might have the potential to grow into something of real interest to the general public.

Chris Muir, Don's widow, says her husband never did things by half measures. "He was very good at his job and he liked working for Canada Cement. He had lots of friends too. We went to company picnics and sing-a-longs and everyone got along. It was the old system, you know, where the company cared for its employees and the employees in turn were loyal to the company. So we had a lot of fun, but Don just loved those birds and that sanctuary. He was the heart and soul of the whole project."

"FATHER" GOOSE

POST CARD FROM THE PAST

I remember many things about my father,

Donald Muir, but especially his physical stature. He stood about 6 foot 2 inches with broad shoulders and an enviable mop of dark curly hair. Stocky, solid but not given to being overweight, he was the kind of man who made walking through crowded Christmas shopping malls easy for my sister JoeAnne and me. We merely tagged along behind his quick steady gait in the inevitable wake created as the crowd parted then rejoined a few feet behind him.

He had steady and confident eyes with large cheeks. His face often widened with a smile and a bellicose and heartfelt laugh. He was often shy when first meeting someone, but grew quickly to a friend and he took delight in learning anything he could from everyone he met. Growing up in Fort Whyte had a profound effect on him as it did on all the residents of the hamlet. In a way, the little town was a world unto itself; a company town within a long walk but years away from its big City neighbour. "The plant" was everything to everyone – employer, sponsor of the hockey team, provider of entertainment facilities and land owner. It seemed everyone worked there – fathers, uncles, brothers, cousins, friends and the punch cards in the small time house at the gate read like the village's school roll call with names like Smith, Kozak, Trenaman, Goodridge, Brick, Burrows, Dupius, Thompson, and Maes.

The plant provided good, well paid manual employment, but it didn't satisfy the intellectual dreams and potential of many Fort Whyte residents including Dad. Like virtually everyone else, he was grateful that the plant gave him a life-long job. But he had a larger vision that tempted him to think beyond the limits of the cement plant and imagine the potential of the site as an outdoor education centre.

Conservation was likely the first long word I learned as a child and it has enduring and special meaning for me, because Dad brought the first geese to Fort Whyte in 1957, the year I was born. From that point on, a good part of our family's life revolved in the glow of my father's dream. In many ways, Fort Whyte was the practical expression of what conservation meant to my father. Dad was for many years a hunter until he felt that habitat loss had reduced many bird populations to levels that he could no longer see hunting as personally reasonable. But he loved those hunting days in the fall marsh as much as he loved those many hours spent caring for the Fort Whyte birds.

He also loved leading groups through the first nature trail in the 1960's and visiting farms throughout Manitoba convincing farmers that their annoying, stubborn little sloughs and low-lying wetlands were better suited for raising wild Canada Geese than grain. Don Muir fundamentally believed in a wise use of resources based on a cautious relationship between the needs of both man and environment.

Recently I accepted a position on the Fort Whyte Foundation's Board of Trustees, and it is like returning home. Thanks to fifty years of contributions from countless volunteers, financial supporters, members and a dedicated professional staff, Fort Whyte has grown into a premier environmental education institution. The early message about waterfowl habitat loss has expanded to include concerns about climate, shrinking habitat, and the woeful inability of young urban kids to access wild places. But the essential theme remains the same. Fort Whyte is a place where I see in every schoolchild's eyes a reflection of a story from a card we received on Dad's death in 1990.

The writer recalled Dad stopping yet another Fort Whyte task, climbing down from a tractor and putting the work on hold so he could excitedly show the visiting children the geese. That moment was typical of Don Muir's enthusiasm for Nature. He honestly believed that if you show a child a goose you can change the world. Reintroduce urban and modern teenagers and adults to the natural world and that experience will guide their future decisions.

Much has changed at Fort Whyte, but much has stayed the same. Fort Whyte is still the place that Don Muir envisioned – a well-germinated seed for a dream that grows everyday, and a place to learn or in many cases relearn how to adopt some "Human Nature".

– Peter Muir

Don's son, Peter Muir in 1967, with a Master Angler Largemouth Bass, 3 lb 2 oz caught at the pumphouse on Pit One

CHAPTER 2

LUCKY GOOSE

It might be hard to believe today, but in the Manitoba of the 1960s, it was a rare and lucky thing to get a close-up view of a Canada goose. And it was unheard of for a flock of wild geese to live and raise their young inside the city.

For many years scientists have debated how many subspecies or "races" of Canada goose occur in North America. Nowadays, most biologists agree there are about seven or eight races. These can roughly be divided into three groups – small geese, (like the Cackling goose), medium sized geese (such as the Moffitt's goose) and big geese, like the Giant Canada. Most of the geese at Fort Whyte are either Interior Canadas, (*Branta canadensis interior,*) which average 3 kilograms in body weight, or Giant Canadas, (*Branta canadensis maxima*), which can weigh as much as 8 kilograms. Giants often interbreed with smaller geese like Interiors so it can be difficult distinguishing them, but in general, any tall, very large goose with a hook-shaped cheek patch and a white spot above its eye is a Giant.

Throughout the first half of the 20th century, habitat destruction and over-hunting had decimated the birds, particularly Giant Canadas, (*Branta canadensis maxima*). Having the southernmost breeding range of all the various races of Canada geese, *maxima* geese were more heavily impacted by hunting than other races of geese. This was due in part to their large size, which worked against them, in that they flew more slowly and responded readily to decoys. Large geese can withstand very cold weather, and Giant Canadas migrated only as far as Minnesota. In the spring, they raised their young in southern Manitoba – in wetlands that had been seriously diminished by agricultural development and the drought conditions of the 1930s. By the 1950s the majestic Giant Canada was feared to be extinct.

As the proud owner of a growing flock of geese and ducks, Don Muir believed that it would be possible to use this flock as the nucleus of a reintroduction program that would restore large geese and maybe even Giant Canadas to the wetlands of southern Manitoba. His plan was to breed Canada geese at the cement plant sanctuary and place them with interested farmers in rural Manitoba, with the aim of both stimulating the wild goose population and encouraging farmers to take more of an interest in the wetlands on their property.

Introducing geese to rural Manitoba would not be as straightforward as it sounds. You can't just raise geese and let them go. Geese have a well-defined system of mating and breeding. A female gosling imprints on its home territory, and she waits until she is sexually mature in two or three years before seeking a mate. Then the male goose, or gander, will follow the female to her breeding territory, which is usually the area or in many cases, the exact wetland upon which she was raised. Geese mate for life, and this further complicates matters. Predators like coyotes and foxes are always looking for a chance to ambush an unwary goose, and if the female is killed, the gander will fly off and look for another mate. All in all, it was a complicated undertaking, and Don Muir realized it was going to require long term funding and numerous other resources.

Alan Scarth, a Winnipeg lawyer, had become friends with Don Muir when they both served as Directors of the Manitoba Wildlife Federation an amalgam of the province's Game & Fish Associations. The Manitoba Wildlife Federation had been formed in the 1940s by Clair Scarth (Alan's father), Dr. Walter Tisdale, and Dick Bonnycastle, the Rhodes Scholar and publisher who later founded the world's largest publisher, Harlequin Books. They all shared an interest in wildlife restoration, a drive that Alan Scarth says was motivated by their experience as hunters: "There is nothing like waiting for dawn in a hide in a Manitoba marsh for engendering long thoughts about the interconnectedness of things. Fort Whyte could not have come into existence if it wasn't for hunters like Don Muir, Sam Fabro, George Richardson, Elliot Rodger, D.I. MacDonald and many others."

Alan Scarth volunteered to call on his connections in the city establishment to ensure that the goose project got the support it needed. As hunters and outdoorsmen, Scarth and Muir wanted not just to reintroduce geese, but to launch a broad conservation effort that would enhance wildlife habitat throughout southern Manitoba.

As Scarth recalls, he and Muir were both worried about the long-term deterioration of wildlife in the province. "I've got a lot of history in Manitoba," says Scarth. "My grandfather founded the town of Virden in 1882, when there were still a few buffalo running around on the Manitoba prairie. He ran a general store but he was a keen hunter and they shot spring and fall, shot everything they could get. Then my father came to Winnipeg in 1909 and studied law and went hunting at every opportunity. They shot huge bags of prairie chickens like there was no end to it. Then came the drought and the Great Depression and my father began getting worried about the declining number of waterfowl and the disappearing habitat. The prairies were drying up and the ducks were disappearing.

"I went into the RAF Coastal Command during the war and when I got back home I met Paul Murphy who flew fighters during the war and became the editor of *The Wildlife Crusader*. Then I met Don Muir, a passionate man and a bow hunter who was likewise interested in conservation. So you had this core group of hunters who really wanted to get involved and reverse this long, grim decline in wildlife populations.

"The trouble was, most of our fellow outdoorsmen at the Game & Fish Association didn't have the long view. They were mostly war vets, working guys who were just happy to be back home and raising families and they thought everything was just fine. They weren't thinking about conservation. They didn't want to join any activist groups and they were very cynical about any endeavour that had a sense of cause to it. So we decided to go outside the Game & Fish Association and create a new organization – one that was seriously dedicated to conserving wildlife and its habitat."

Scarth assembled a group of prominent community members who could provide guidance and help raise funds for the goose program and other worthwhile endeavours. The founding trustees were Donald Muir, Stewart Perdue, George Richardson, Elliot Rodger, Alan Scarth, Clarence Tillenius and George Tritschler. (They were subsequently joined by Jennifer Shay, Harry Duckworth, Sam Fabro, and Mac Runciman.) On April 27, 1966, the Manitoba Legislature passed a special act that incorporated their new group, that they called "The Wildlife Foundation of Manitoba."

Scarth says, "We knew people would ask us, 'What's so special about your organization?' So we came up with an explanation:

'Over 90% of the wildlife in southern Manitoba lives on private land. The future of wildlife is therefore in the hands of private landowners. The Foundation will go to them and suggest that they set aside wild lands. The landowners will be offered a wide range of assistance. The Foundation will provide breeding stock, offer advice on improvement of water and upland habitat, and give access to technical engineering services. But the landowners will do the job themselves, and they will do it enthusiastically and with more efficiency than any other agency. That is what's different about the Foundation.'"

NATURE NOTE

he most productive region on the planet for waterfowl breeding is the fertile rairie Pothole region of Western North America, encompassing parts of Manitoba, askatchewan, Montana, North Dakota and South Dakota. Over several decades, this vital abitat has been reduced by agricultural drainage, but is increasingly being preserved ow that we understand its importance for wildlife, water quality, and flood control.

The Wildlife Foundation of Manitoba started with a small trust fund – a few thousand dollars – but with serious energy. Don Muir, the manager, hit the road and introduced himself and his organization to farmers all across the southern fringe of the Province. It was a critical time for wildlife because habitat loss was accelerating at a frightening pace. Wheat prices were at an all-time high and farmers, equipped with expensive and powerful new all-wheel drive tractors, were clearing brush and cultivating their land right down to the water's edge. In dry years, some farmers were ploughing right through shallow sloughs and coaxing a crop out of low ground that most years would hold water. Even in wet years those low spots would thereafter be devoid of vegetation and useless to wildlife. The tax laws favoured this kind of habitat destruction, and in fact farmers were penalized financially if they didn't indulge in it. So wetlands – the province's most productive wildlife areas – were disappearing rapidly, and with them innumerable birds, animals, and amphibians.

Alan Scarth visiting partners in Southwestern Manitoba

Despite the environmental problems associated with high-intensity agriculture, farmers tend to be outdoorsy people who appreciate nature more than the average city person. Farmers live on the land, and they are keen observers of weather, rainfall, wildlife populations and other natural cycles. So even though the post-war agriculture industry was inflicting a lot of damage on the landscape, there was a certain amount of sympathy amongst individual farmers for the case the Wildlife Foundation was putting forward.

Don Muir was a tall friendly man with a knack for connecting with people. Many farmers were concerned themselves about the destruction that new technologies and antiquated tax laws were causing, and in some cases it seemed like they'd been waiting for someone like Don to show up in their driveways. "The land owners are on our side," Don reported after a tour of rural Manitoba. "In each community I can pretty well count on landowners willing to set aside wildlife areas under our program. Our only limitations are time, money and breeding stock."

Wildlife foundation

A bill to incorporate a private foundation working for wildlife conservation was approved in principle in the Legislature Friday.

Introducing the bill, James Bilton (PC — Swan River), said Manitoba and western Canada faced a rapid depletion of wildlife.

"The bill provides for a group of well known private citizens to band together to do what they can to halt the trend," he said.

The foundation is a non-profit organization which will channel funds and support to assist established government agencies in the field, he said.

The bill lists as members Donald W. Muir, H. Stewart Perdue, George T. Richardson, N. Elliot Rodger, Alan Acarth, Clarence Tillenius, and George E. Tritschler (Queen's Bench chief justice).

March 26, 1966, Winnipeg Free Press

Alan W. Scarth Q.C., founding Chairman of The Wildlife Foundation of Manitoba - established by a Special Act of the Manitoba Legislature in April 1966

Muir, like many of the founders of the Foundation, was a careful man who lived by the old tradesman's adage "Measure twice before you cut." The last thing he wanted was to antagonize landowners or do a half-baked job in some community and create a ripple of gossip that would spread from town to town. So he proceeded methodically, moving forward step by step, avoiding commitments that the Wildlife Foundation wasn't able to handle. "My only concern is that we keep our growth within bounds," he wrote. "We need to be sure of providing the essential minimum of educational input and guidance for each project."

Alan Scarth and Don Muir made up a batch of durable metal signs and offered them to farmers to mark their wetlands projects. (Some are still in place to this day.) Then Scarth and the other Foundation Trustees worked on the urban front. After all, not all the good habitat in the southern part of the province was out in the hinterland. Some of the best habitat was along the Red and Assiniboine Rivers. Riverbank landowners, then as now, tended to be affluent, well-educated and nature-oriented people, so they seemed like a good target group for the Wildlife Foundation to pursue. But the Wildlife Foundation needed a symbolic species to carry their message – a Manitoba version of Smokey the Bear.

Releasing Fort Whyte geese on a pothole in Southwestern Manitoba

Like Giant Canada geese, wood ducks were quite rare at the time, largely because the old, dead trees that hold their essential nesting cavities had either been cut down by landowners or inhabited by raccoons – a species that is pure hell on wood duck eggs and, like white-tailed deer, has spread through Manitoba in synch with the growth of the human population.

Branta Newsletter, 1997

OBITUARY

• CHESTER •

Faithful Companion & Media Star

Possibly the best-known dog in Winnipeg passed on this summer. Ken Cudmore's famous Yellow Labrador, Chester, died 11 August, 1997, from cancer. Chester worked at FWC one day a year - every fall, when Ken brought our resident flock into the Waterfowl Room for the winter. Chester dilligently rounded up the waterfowl, ignoring the icy water. He would gently carry the birds, uninjured, back to Ken. Over the years, Chester's annual duty was a regular media event. Chester will be missed by us all.

Fall Roundup: Ken Cudmore's soft-mouthed Yellow Labrador Retriever, Chester, was an invaluable partner for rounding up the less cooperative members of the display flock for their 'migration' indoors for the winter

"If you build it, they will come."

The trustees of the Wildlife Foundation selected the colourful wood duck because it symbolized the importance of habitat....and a habitat literally right in the backyards of city dwellers. Today, most people understand that habitat loss, not hunting, is the main factor in decreasing wildlife populations. But in the 1960s and early 1970s people didn't really understand the importance of habitat. (Years later, the Canadian writer W.P. Kinsella coined an epigram that could have served as a perfect motto for the Wildlife Foundation of Manitoba – "If you build it, they will come.")

Scarth says that making people understand the relationship between habitat and healthy wildlife populations was strictly a matter of education. "We needed to teach people about the significance of habitat," he says. "So we came up with a slogan – 'No Habitat, No Wildlife.' We felt that these four words compressed a whole century's worth of scientific research. If we encouraged urban landowners to install nesting boxes on their river-banks, it would drive home the message about habitat, and it would perhaps encourage them to get involved in larger conservation issues."

Joint annual meeting of the Boards of Trustees and Management, circa 1986

The trustees decided to launch an intensive wood duck reintroduction program, using the Fort Whyte waterfowl sanctuary as a place to house and maintain the birds until they could be released in suitable places around the city. One of the founding members of the Wildlife Foundation, Justice George Tritschler, donated the necessary funds to purchase 50 wood ducks from Guelph, Ontario, and have them delivered to Fort Whyte. The idea was to build wood duck nest boxes, install them in good spots along the riverbanks, then release wood ducks near the nesting sites in the hope that they would imprint on that area and return the following year to raise their young. It seemed quite workable on paper. But the best laid plans of mice and men...

The Ontario ducks arrived at Fort Whyte and were installed in the waterfowl house, where they were kept under the watchful eye of Don Muir. It was early spring, and there was still snow on the ground. The ice on Lake One was still solid but there was a small patch where water returning from the building had opened a hole in the ice. This patch of open water looked harmless enough, but it would prove to be the doom of the newly arrived wood ducks.

Mink

NATURE NOTE

Next to otters, mink are the most aquatic members of the weasel family, or mustelids. The feet of a mink are partially webbed for fast, easy swimming, and its tail is used as a rudder.

Charcoal sketch by Clarence Tillenius

One day, not long after the ducks arrived, a mink came loping down the shoreline. The mink's nose no doubt crinkled with pleasure when it detected the tantalizing smell of warm duck wafting on the breeze. The mink quickly got to work inspecting the building, sniffing at every nook and crevice. If the mink could somehow wiggle his way into the waterfowl building, it would have an epic feast. The building was of course sealed tight, but the mink's visit did not go unnoticed by Don Muir, who spotted the animal's dainty tracks in the snow. Don Muir and the mink soon became silent adversaries. Every day, Don spotted the mink's tracks as it came back to check the doors and walls of the building. Every day, Muir was relieved to see that the mink left after determining that entry was impossible. Don was uneasy about the mink's persistent visits. A mink can squeeze through a tiny hole and burrow and climb if necessary. But he had been keeping birds in the old building for several years, and so far the defences had held.

Piping in the first Canada Geese, CBC Breakfast with the Birds, March 1988

One morning Don walked over to the building and noticed there wasn't a single mink track around the building. He assumed that the mink had finally given up. But as he pushed the door open, his heart sank. Normally, the wood ducks roosted on the cross members of the building. But only two ducks were up there, and none were in the shallow pond that covered the floor. Looking around, he saw that a nesting box in the corner was stuffed with dead ducks. The true scale of the slaughter wasn't evident until he looked into the darkest corner of the building. There he found all the other dead ducks neatly stacked, each with a bite to the back of its head.

After checking the inside walls of the building, Don could not figure how the mink had gotten in. Then he noticed that the drain cover had been knocked off the stand-pipe. Suspicious, he went outside and carefully checked the surface around the edge of the open water. Sure enough, mink tracks! The mink, with an incredible sense of smell, had found the underwater outlet, carrying the aroma of duck, and like a fearless daredevil, had wriggled his way up the long pipe. One can only imagine the pandemonium that broke out in the waterfowl building when the mink's fierce little visage emerged from the standpipe. After finishing its killing spree, the mink had left by the same route. And Don Muir might have never sorted out the mystery if he hadn't noticed the drain cover ajar.

THE KILLING SPREE

The old expression "weasel in a hen house" comes from the well-documented habit of weasels, mink, and other members of the order *mustelidae* to keep on killing penned-up animals until they are all dead.

There's an old fallacy that only humans hunt for sport. But many animals will at times kill other animals for reasons that have nothing to do with food. Felines often stalk and kill birds for entertainment. Wolves tend to kill dogs whenever they get the opportunity. Large male bears will kill young cubs if they can get past the mother. And owls swoop down on roosted crows at night and dispatch them with a quick squeeze of the claws – perhaps in revenge for the crow's habit of harassing owls in the daytime, which in turn might be payback for the owl's habit of harassing crows at night. But the mustelids are particularly notorious for killing like there's no tomorrow.

When a mink or weasel gets the opportunity to kill a large number of birds or other animals, it usually caches the bodies with orderly care, intending to perhaps return later when food isn't so plentiful. We can condemn this tendency to hoard food, but it many ways it beats subsistence hunting as a survival strategy. A species known as *Homo sapiens* developed this same habit about ten thousand years ago – a period that anthropologists call the Dawn of Civilization.

Don could have wired the cover down. But this was one smart mink, and now that it had acquired a taste for captive duck it would never be deterred. So with some reluctance he set a trap alongside the building, and that night, the perpetrator's duck-hunting career ended with a snap.

After Muir had killed the mink, he showed the carcass to Alan Scarth and related the sad story of the murdered ducks. Scarth in turn took the frozen mink to Clarence Tillenius' old studio at the foot of Dominion Street where the renowned wildlife artist sketched the animal with a piece of charcoal.

That might have been the end of the fledgling wood duck program if not for the generosity of George Tritschler, who expressed admiration for the mink's ingenuity and gallantly offered to pay for another 50 wood ducks. This time, the waterfowl house held firm against the assault of predators and the ducks survived.

Come summer, the "woodies" were released at prime sites alongside the river. Over the following winter, Jim Ashdown and Elliot Rodger made dozens of wood duck boxes in Jim's basement. And in spring, Elliot Rodger, the former Major-General and self-appointed "Chief Box Inspector" tramped through the spring snow with his trusty sidekick Des Whellams and climbed rickety ladders to put them up in selected trees. When the wood ducks returned from their wintering grounds in the deep south they found spanking new homes waiting for them. Every year, Rodger and his partners cleaned the boxes, reasoning that no self-respecting mother duck wants to raise her youngsters in a messy home.

THE WOOD DUCK

Most people would probably guess that the mallard is the most common duck in the Winnipeg vicinity. But in fact it's the wood duck that is our most common nester. In the late 1800s, wood ducks were hunted to the brink of extinction. But tightened laws and vigorous conservation efforts have helped this bird recover to the point of being one of our most successful species.

The "woodie" might be the prettiest duck in the world. With its resplendent plumage and handsome crest the male wood duck has a dash of almost every colour in the palette. But it's also a reclusive bird and many Winnipeggers have never seen one. The woodie's preferred habitat is the fringe of hardwoods along the city's riverbanks. On a summer day, if you watch closely, you'll see this small duck paddling down the river close to the shore. They are quite comfortable on land and if a boat approaches they will sometimes scurry up into the woods.

Unlike most ducks, woodies nest in tree cavities, and Fort Whyte's nest box program has been a tremendous aid to their recovery. In the spring, hen wood ducks return to the neighbourhood of their birth, bringing their mate along with them. (Drakes have no concept of "home," and they often follow their mate hundreds of miles from where they were raised.) In natural circumstances, hens will nest in tree cavities. But these are not always easy to find, and are sometimes occupied by raccoons or kestrels. So the hen's chances of raising a family are much improved by the ample availability of sturdy, clean nest boxes.

Incubation lasts for about 28 days. When the ducklings emerge from their eggs, the mother coaxes them to jump down from the nest. They are covered with fluffy down and weigh so little that they can fall 50 feet without injury. They follow their mother to the river, where they spend the summer foraging for aquatic insects and plants. In autumn, these young ducks will fly south and link up with mates of their own, and the females will bring their partners back to Winnipeg. If they are lucky, they will then establish a home of their own – in one of the 200 nesting boxes that Fort Whyte volunteers maintain within the city limits.

Elliot Rodger

The Winnipeg Wood Duck Project soon became a topic of much interest amongst the riverbank homeowners of Winnipeg. Keeping up with the Joneses may have once required the acquisition of a new garden or patio deck, but now it included having your very own resident family of handsome wood ducks. The trustees of the Wildlife Foundation believed that if homeowners installed a wood duck nesting box on their riverbank property, they might be so thrilled by the sight of these avian dandies in their backyards that they might volunteer their time and money to the Wildlife Foundation. This in fact happened in some cases.

Fran Pollard, for one, first became interested in wildlife conservation when Elliot Rodger installed a wood duck box on her riverbank property. The box captured the interest of some wood ducks, and Fran Pollard was in turn captivated by the ducks. She was caught up in the cause of nature conservation and signed up to work as a Volunteer Interpreter leading school groups, and went on to become one of Fort Whyte's longest serving Volunteer Interpreters.

In many ways, the wood duck program turned out to be one of Fort Whyte's greatest successes. From a couple of nesting boxes, the program grew steadily until over 100 boxes were installed by 1980, most of which successfully produced families of wood ducks. Elliot Rodger spent 12 years risking his neck climbing rickety ladders for the sake of wood ducks, then retired to Ottawa in 1982, at which point the job of "Chief Box Inspector" fell upon a group of volunteers who determined that they would try and fill the Major-General's shoes.

The assorted volunteers built new boxes, made contact with new home owners, and by 1990, the wood duck population of Winnipeg had more than doubled, a remarkable fact given that during the 1960s, the mere sighting of a wood duck would draw bird watchers from around the city. Now, some home owners reported congregations of 30 wood ducks foraging for acorns in their backyards, and Conservation Officers reported large flocks along the Assiniboine River in Headingley. And most of these birds probably originated from the small flock raised at Fort Whyte.

Today 100 wood duck boxes along the City riverbanks continue to receive annual spring maintenance and nesting activity reports by Fort Whyte volunteers

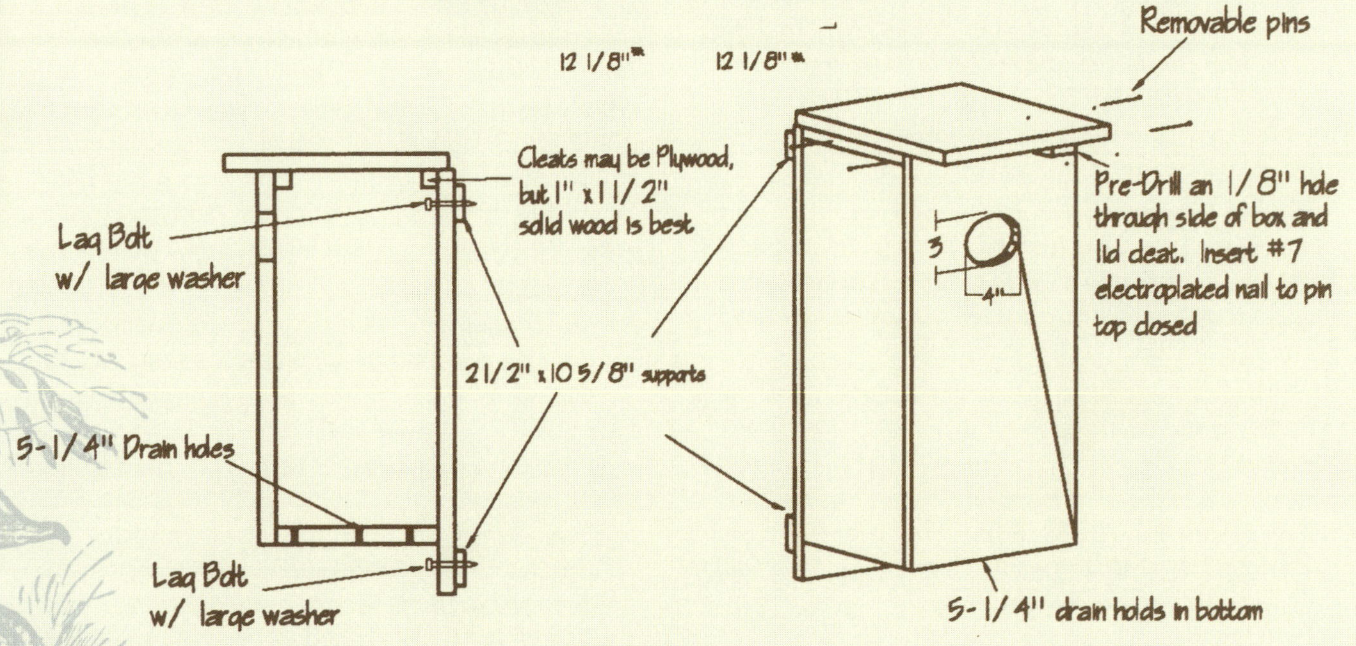

Cubs and Scouts have been regular visitors to Fort Whyte since the early '70's

With successes like the wood duck program under its belt, The Wildlife Foundation of Manitoba began gaining credibility in certain quarters of the government and the business community, and corporate donations from companies like Monarch Life and Hudson Bay Mining and Smelting began trickling in. The central challenge, in many ways, was to get the message across to everyday private citizens. Muir was a hands-on guy, accustomed to tackling problems and solving them on his own, and he didn't have much trouble speaking a language that business people and farmers understood. But the "environmental movement" hadn't been born yet, and city people weren't much concerned about far-removed issues like habitat loss in rural Manitoba.

When the Wildlife Foundation got started, Scarth had been frustrated by lack of conservationist zeal in the fishing and hunting community, but he believes that non-hunting, non-angling urbanites turned out to be worse. "They were much worse," says Scarth. "In those days, your city dweller didn't care in the slightest about vanishing wetlands. They thought wetlands were where mosquitoes came from, and good riddance."

Scarth and Muir and the other members of the Wildlife Foundation had spent a good ten years trying to rally the support of educated, affluent city people, and they finally admitted they were getting nowhere. "We decided we had to write them off as a lost generation," says Scarth. "We had to try a different tack."

They theorized that they might make better headway by taking their message to young people. "We called it the 'trickle up' theory," says Scarth. "We theorized that the kids would catch the conservationist bug and pass it on to their parents. At the very least it would be time well spent because they would eventually grow up and become the legislators of the future."

The best place to find large captive groups of young people is of course the school system, so that's where the Wildlife Foundation directed its efforts. On an experimental basis, the Foundation organized some field trips, bussing high school science classes from Brandon and Winnipeg to the Delta Marsh, where Professor Jennifer Shay managed the University of Manitoba's Research Station. With the help of other professors she tried to teach the students about conservation principles. The approach worked well with older high school students, but not so well with elementary school kids. "It was a long bus ride," says Scarth. "And when you put thirty-five younger kids on a bus they tend to get a little hairy on you."

After a number of these elementary school trips, it became apparent that the wildlife at the Delta Marsh would never compete with the wildlife on the bus. City kids, raised on television and smart-alecky sit-com repartee, seemed particularly hard to captivate. But the problem was perhaps not so much the message but the medium. "Youngsters aren't really built for sitting still on a bus for ninety minutes and then listening to adults drone on and on about Nature," said one teacher. "By the time we got to the Delta the boys just wanted to get out and run around and scuffle with each other and flirt with the girls. And who could blame them? They were just acting like the ducks and geese we wanted them to study."

It was obvious that the Foundation needed a spot close to the city. It needed to have wildlife but also be sufficiently self-contained that the teachers could keep an eye on the kids and monitor their behaviour and safety. One bad incident could damage the program permanently. Scarth and Muir examined an old hunting lodge at Libau, which was at least closer to the city, but it was rotting and falling down. One day in the spring of 1972 Scarth and Muir were out at the Fort Whyte waterfowl sanctuary pondering this dilemma when a light bulb flashed on.

As Scarth recalls, "Don and I looked at each other and said, 'Wait a minute, what's wrong with this place?'"

Indeed, Fort Whyte seemed made to order, but there was one slight glitch, the property belonged to Canada Cement. And Canada Cement was in the business of making money, not educating children. The company had tolerated the presence of the waterfowl sanctuary, and occasional groups of Boy Scouts, Girl Guides and student visiting the property to see the birds and walk the nature trail, but Don Muir felt that Superintendent Harold Edmonds had been stretched to his limits on that front. The Foundation wanted to bring in thousands of kids, not hundreds, and it was highly unlikely that the company would tolerate the insurance liability issues, the potential damage to property, increased costs of fencing, signage, road maintenance, garbage removal and so on. The whole thing was inarguably a massive potential headache, and what could they offer the company in return? Certainly not money, they didn't have any.

Alan Scarth realized this would be a delicate but extremely important part of the Foundation's future, so he stuffed his lawyer's valise with documents, knotted his tie, and went off to see Superintendent Edmonds.

Scarth was well aware that the super was a formidable man. "I knew ahead of time that he hated environmentalists. Apparently in his mind they were akin to terrorists. That might sound rather comical today but you have to remember that this was a very different time. It's hard to believe now, but environmentalists were considered to be real extremists in those days. Many people in positions of power were still in love with the idea that a wide-open marketplace and untrammelled economic growth were the cures for society's ills. Even the provincial government's recent slogan was 'Growing to beat 70.' The Foundation was preaching a very different message – that careless industrial growth was seriously damaging our environment and jeopardizing its future health, and that school children needed to be made aware of this. Canada Cement was one of our leading examples of old-style smokestack industry. And in retrospect, I think we were really asking a lot to expect that they would take our side in a project that might cast a negative light on their own activities."

Straw bales: a simple approach to encourage goose nesting

Ring-necked Duck

NATURE NOTE

Though most diving ducks prefer deep, open water, the "ring-necks" are common on small prairie ponds during migration, and nest primarily in the boreal region of North America. It is most easily recognized not for its namesake faint purple collar, but by the striking white ring found encircling its bill.

So to say the least, Scarth anticipated that the meeting would be a tense one. The Superintendent's office was on the second floor of a nondescript grey building, now demolished, on the north side of McGillivray Boulevard. Scarth endured a long wait in the outer office then the superintendent's secretary led him into the inner sanctum. "This large, godlike figure of severe countenance sat behind a wide desk, glaring at me," recalls Scarth. "He did not arise from his chair to shake my hand and he did not invite me to sit down. He just nodded imperiously and said, 'You have two minutes.'"

It wasn't a promising beginning. But Scarth, being a lawyer, had prepared himself for the worst. Rather than allow Harold Edmonds to set the ground rules, he politely folded his briefcase and cancelled the meeting before it started. "I replied that I couldn't say what I needed to say in two minutes so there was no point in wasting his time and mine."

Scarth reasoned that if he gave Harold Edmonds the opportunity to say no it would be all the more difficult for the man to swallow his dignity and reverse the decision at a later date. So he took his leave and decided to go over Edmonds.

The company's manager for western Canada was a man named John Redfern. "I decided to take a chance and approach him," recalls Scarth.

Scarth hadn't lost hope that he could make a deal. Canada Cement was a company that prided itself on being a good corporate citizen, particularly when it came to supporting initiatives that its employees were involved in. He hoped that Don Muir's involvement as a valued employee meant that Redfern might at least give the Wildlife Foundation proposal a fair hearing.

Nowadays, John Redfern lives in Montreal, where he still works for the company that took over Canada Cement, LaFarge Canada. Redfern describes that initial meeting with Scarth with some amusement. "Alan is a very persuasive guy," he says. "He's always bubbling with ideas, and he no sooner ties up one idea than he's moving on to the next. He happened to be approaching us at a very good time. We were becoming quite conscious of our vulnerability as an industry. We are, after all, a company that gives off airborne effluent, and our plants tend to be quite conspicuous. Think of the plant outside Banff, for example, alongside the Trans-Canada highway. Here you have these beautiful mountains, and in the foreground a tall chimney pumping smoke into the clear mountain air. So I could already see how those smokestacks might become a public relations problem for us in the future."

Canada Cement is an old company (est. 1911) and it got its start in times when smokestack effluent was regarded as not only a tolerable by-product of industry but something to point at with pride. "It was a sign of prosperity," John Redfern remarks wryly. "If there wasn't smoke coming from the chimney it meant we weren't making any money. In fact our old stock certificates were illustrated with a chimney belching smoke."

Paul Murphy presents Sam Fabro with cheque from Labatt Game & Fish

PAUL MURPHY

One of the most colourful characters at Fort Whyte was Paul Murphy. With his silvery coiffure, hearty laugh and movie-star good looks, Murphy made an impression wherever he went. But Paul Murphy was more than just a handsome face. An outstanding athlete as a young man, Murphy was a feisty competitor who flew Hurricane fighters in the Second World War and developed a taste for risk that stayed with him all his life.

He was perhaps best known in Manitoba as the plain-talking, persuasive editor of the ground-breaking conservation magazine *The Wildlife Crusader.* But he also developed a national reputation when he rallied sportsmen across the country and founded the Canadian Wildlife Federation. After editing *The Wildlife Crusader* for 30 years, Murphy turned his attention to promoting the environmental education program at Fort Whyte. He might have been a tireless conservationist and outdoorsman but he also loved the night life and never missed a party. After his retirement he decided to make a six-figure donation to Fort Whyte. Alan Scarth recalls that the donation caused quite a stir. "The natural question in all our minds was, whence came this largesse? The Federation had never paid him more than a modest stipend. Where could he have got the money?"

Paul Murphy wrote a book about his life that answers that question. In his memoir *A Gambler in War and Peace*, he reveals that he made a very small salary at magazine ($10,000 per year) but he often flew gambling tours to Las Vegas, and was one of the few people on the plane to win regularly. "I made money," he recalls. "Some may scoff but I have proof. I paid for a nice house in Winnipeg, condos in both Winnipeg and Phoenix and always lived comfortably. But I was always generous. I firmly believed that my continued good fortune was dependent on giving some away."

Murphy gave more than just his time and money to Fort Whyte. His real gift was *joie de vivre.* Heather Anderson, a veteran Fort Whyte receptionist, says Murphy often showed up with flowers and treats for those of us who ran the office. "He never needed an excuse to celebrate," she says. "He brightened up your day just by walking in the door."

Scarth gave Redfern his very best spiel. "I told him that I could understand why he might be hesitant to introduce wet-behind-the-ears environmentalists to his property but it might be good for his company. I said, 'How about leasing us some land along McGillivray and we'll put up our own buildings. If things don't work out in ten years you can kick us off.'"

Redfern listened carefully to Scarth's pitch and agreed to discuss it with the other senior managers of the company. "I thought it was a good idea," says Redfern. "Don Muir and the other people from the Wildlife Foundation had done a very good job at Fort Whyte with the bird sanctuary, so I was favourably inclined towards his proposal. To me, it was a win-win situation. Alan is very generous about praising me for the role I played in getting them a lease but it wasn't like I had to go to the company on my hands and knees. It seemed like a perfect project for us. They would handle all the expense and the work and we would get credit for being involved in something that seemed very worthwhile."

Scarth insists that John Redfern is too modest about his role in getting the lease approved. "There's no limit to what you can achieve in life if you're willing to give the credit to someone else," he says. "But the fact remains that if it wasn't for John Redfern, Fort Whyte wouldn't exist."

With John Redfern's assistance, the Wildlife Foundation signed a 10-year lease with Canada Cement, (for one dollar) and embarked on a serious campaign to deliver environmental education to young people.

CHAPTER 3

LIGHT ON OUR FEET

It was obvious that before the program could proceed there were many details to take care of. After all, where would all these kids go to the bathroom? What if it rained? Who would clean up the chewing gum wrappers and other garbage that would inevitably appear like magic when hundreds of kids walked through the woods every week? A central reception building of some kind was also clearly needed. But buildings cost money, and the Wildlife Foundation was still puttering along with tiny grants and whatever its members contributed out of their own pockets. Most of their infrastructure had been scrounged. In one case, Chief Wood Duck Box Inspector Elliot Rodger had liberated a metal shack from the Air Force base in MacDonald, Manitoba, where it had been used as a residence during the war for the Commonwealth Air Training Plan. Rodger was no longer active in the military, but he retained his old connections, and he used them to get a "Fort Whyte Discount" on the building, which he then trucked, with much ceremony, to Fort Whyte, where it was put to use as a brooder and wintering facility. But they couldn't get away with using another weather-beaten Air Force building as a base of operations for professional school tours. Nor could they go back to Canada Cement and ask for money to put up a building. The company had already contributed more than its share of fencing materials, bird feed and of course the property, so it couldn't be asked to kick in a building too.

It was estimated that a basic reception building would cost at least $25,000, and in the Winnipeg of 1972 that was a lot of money. The whole enterprise might have languished at this stage of its growth if not for the intercession of a colourful individual named Sam Fabro.

A successful Winnipeg businessman, Sam didn't have an enemy or even a detractor in the entire city. He had a smile for everybody, the laconic shrug of a man who has seen everything, and the trademark broken nose of an old-time hockey player. He was also a hunter and a member of the Game & Fish Association, where he had forged strong friendships with Don Muir, Alan Scarth and the other founders of the Fort Whyte sanctuary. "Sam was the perfect team mate for me because he's a people person," says Alan Scarth. "Sam gets along with everybody."

In 1969, Scarth (who for some reason everyone calls by his last name) telephoned Sam (who for some reason everyone calls by his first name) and invited him out to Fort Whyte to do some volunteer work. "It looked like a moonscape to me," says Sam. "Not pretty at all. But Scarth and I got to work digging postholes for a predator proof fence."

Once Scarth succeeded in getting Sam on the property he invited him to join the Wildlife Foundation. In many ways Sam was an ideal draftee. He loved being outdoors, had helped organize the 1967 Pan Am Games and was a leader of the local Kiwanis Club of Winnipeg. More to the point, he was interested in working for the conservation cause. "I was a keen duck hunter," he says. "I used to shoot up at Twin Oaks Lodge on the Netley Marsh. The marsh isn't any good anymore. It's been ruined by high water and invasive species, but it was beautiful in those days. My family is from Italy, and I've travelled quite a bit, but you'd have to go a long way around the world to find country prettier than the marshes of this Province in autumn. The smell of the water and the huge prairie sky and the sight of all those migrating birds – it's really terrific. Those fall hunting trips on the marsh are some of my greatest memories of living in Manitoba."

Like Alan Scarth and Don Muir, Sam could see the landscape declining, and was worried that the younger generation would never have the chance to enjoy the outdoors as he did. "Almost all the fellows who started Fort Whyte were hunters," he remarks. "I guess hunting isn't as popular as it used to be. And that's a shame, because most of the protected habitat we have in Canada today wouldn't exist if your everyday hunter hadn't opened his wallet."

Bud Malcom

He says the Kiwanis Club of Winnipeg was always looking for worthwhile projects to support. And since he was Lieutenant Governor of the Kiwanis at the time, he didn't have much trouble steering the Club in the direction of Fort Whyte. "Exposing school kids to nature and teaching them about conservation looked like a pretty worthwhile undertaking. So we decided to try to raise some money and support it."

Every year the Kiwanis Clubs organized 10 television auctions. In 1973, the proceeds for all 10 auctions were given to the Fort Whyte Nature Centre – $25,000, which coincidentally was exactly the Foundation's budget for the new building. Construction on the building started later on that year, with well-known Winnipeg contractor Bud Malcom (Malcom Construction) handling the job. Malcom Construction was a family-run business with a good reputation all around the province, and its patriarch Bud, a *mensch* of the first order, took on the Fort Whyte project out of sheer charity. "It wasn't the easiest project for him," says Sam. "The poor guy had a lot of trouble sinking piles in that quicksand out there."

The plan was to mount the building about twelve feet above grade to enhance the view out over the lake and marsh. Malcom attempted to accomplish this by pouring a number of concrete stilts to support the beams and understructure. But the very clay that made the quarries valuable to Canada Cement expanded and contracted with moisture and frost, leaning the pillars with every change in temperature. Bud Malcom eventually solved the problem, and in 1974 a group of supporters gathered to celebrate the launch of the new Kiwanis Reception Building. As Sam says, "It was significant milestone in the evolution of Fort Whyte."

The Winnipeg Foundation followed up by donating $22,000 to construct a new waterfowl building, in which romantically entangled pairs of geese were given their own 50 square foot apartments with 100 square foot swimming pools. This gift was enabled by Alan Howison, who was then the CEO of the Winnipeg Foundation and, in the words of Alan Scarth, "had the sensitivity, imagination and courage to understand the Fort Whyte vision at a time when environmental education was new and little understood. His support established a connection with the Winnipeg Foundation which has proven, over time, to be one of our major contributors."

The waterfowl building had an incubator room and a classroom from which students could observe the avian families through large glass windows. The University of Manitoba designed a series of study programs that would give students credits for doing research at the new facility. Once it was up and running, the new waterfowl building produced about 200 goslings a year, which were distributed to prime habitat locations throughout southern Manitoba.

Waterfowl pen construction, 1974

The newly constructed Kiwanis Reception Centre overlooking its waterfowl rearing and display pens

"Much of the credit for our survival during that period belongs to Sam Fabro," says Alan Scarth. "His Management Committee repeatedly accomplished the impossible during our early years and he somehow saw to it that we escaped the fate of the Winnipeg Symphony, the Opera, and the other admirable non-profits that ran out of money and had to go begging to stay alive. It's one thing to raise the money for a building, and quite another to come up with the operating costs. These ongoing costs are the downfall of many a non-profit, and in fact there's an old saying in the real estate trade: "If you really dislike someone, give him a building." And it wasn't like Sam didn't have other responsibilities in his life. He was CEO of a major floor covering distributor, and was running a long list of charities, but he somehow made the time to organize scores of people and do all our public relations work. If not for Sam Fabro's natural flair for getting people behind a cause, Fort Whyte would not be where it is today."

Throughout the 1970s the Fort Whyte Centre used the new facility as a headquarters for educating students in the ways of nature. Periodically, other donations came in from diverse and sometimes surprising sources. It was not unusual for people to "discover" Fort Whyte for some coincidental reason, then get into the habit of coming out for walks in their spare time, acquire a feeling of ownership, then decide to either volunteer their time or donate some money to the cause.

Evening Grosbeak

Many of the donations to the Fort Whyte Centre had their origins in just such random conjunctions of events. In the 1970s, for example, one of the most popular species at Manitoba bird feeders was a colourful, noisy, gregarious bird called the evening grosbeak.

Evening grosbeaks were very common in Winnipeg in the 1970s, and Dorothy Johnston had a regular flock of grosbeaks visiting the feeder in her backyard. She was a widow, whose son was posted to a far away RCMP detachment. She loved nature and the birds kept her company through the winter. Evening grosbeaks were her favourite, and she enjoyed watching them cracking open the sunflower seeds with their powerful bills and extracting the meat with one efficient motion.

One of Charles Darwin's most important contributions to science was in emphasizing the importance of random chance in Nature. In western philosophy and religion, we tend to emphasize the importance of design and purpose in the shaping of destiny. But if we examine the most important features of our personal histories – how we met our spouses for example, or how we ended up in our careers - those grand events can usually be tracked back to what poets call "fate," and what scientists call chaos or random coincidence. This is just one of the many ways that the observation of nature offers insight into our own lives.

Rose-breasted Grosbeak

NATURE NOTE

Grosbeaks are the largest members of the finch family. The Rose-breasted Grosbeak, the pink-hued Pine Grosbeak, and the yellow-and-black Evening Grosbeak call Manitoba home. The Evening Grosbeak is an important natural control of the Spruce Budworm population that ravages important timber resources throughout Canada's boreal forest region.

Like all bird enthusiasts with a feeder on the go, Dorothy needed a constant supply of sunflower seeds, and she heard that Alan Scarth's wife Helen had a garage full of seed at her home on North Drive in Fort Garry. Living as she did in a woodsy neighbourhood near the Wildwood Golf Course, Helen Scarth had to feed several hundred hungry grosbeaks every day, so she obtained her bird feed in massive 500 pound lots from a Fort Whyte regular named Des Whellams. (Whellams is another unsung hero in the Fort Whyte story. He was not only Elliot Rodger's sidekick in the wood duck program, but was, according to Scarth, "a meat hunter for Winnipeg butchers as a young man, who then converted and became a conservationist with the Winnipeg Game and Fish and the Manitoba Wildlife Federation. Late in life he was our dedicated supporter and our first major endowment funder. He handed the key to four year-old Billy Elliott when Carl Nelson's new building was opened, and left us his entire estate of $250,000 in his Will.")

Helen Scarth thusly became the go-to person for bird feed in the neighbourhood. And when Dorothy ran short, she would walk across the street to pick up a fifty pound bag from Helen, stopping for a chat over coffee. Because of that short-lived abundance of grosbeaks in Winnipeg, Dorothy Johnston and Helen Scarth became close friends.

In those days, Helen Scarth was an active partner in the development of Fort Whyte. She held the surveyor's rod when the first building on McGillivray was being sited, and knew much about her husband Alan's role in the construction of the early buildings and Canada goose pens. So it was natural that Helen and Dorothy talked about Fort Whyte over coffee. Still, when Dorothy eventually passed away, it came as a surprise when she left her estate to the Fort Whyte Foundation. "Capital money was in very short supply in those days," says Alan Scarth. "And we were very grateful."

A cairn acknowledging Dorothy Johnston's gift was erected along the walk to the marshland, but not many of the regulars who walk past it every day are aware that Dorothy Johnston's bygone flocks of evening grosbeaks share much of the credit for her gift to Fort Whyte.

Wayne Neilly and kids

During the 1970s, the Centre hired a manager, an accomplished and dedicated naturalist named Wayne Neilly, who worked with teachers and curriculum designers from the provincial education branch to make sure that the Centre's nature programs were a good fit with the school curriculum. The flow of ideas went back and forth. In most cases Neilly and the rest of the Fort Whyte team designed programs that were specifically built around course material. Students would study a topic in the classroom (like how a species adapts to a particular environment,) then they would go out to Fort Whyte and take a walk through the woods and look for examples of how adaptation works in the real world. In other cases, Fort Whyte came up with the ideas. As a private organization, Fort Whyte had the agility to move forward without getting tangled in bureaucratic red tape, and sometimes these program innovations simply sprang from some volunteer remarking, "Hey, what if we did it this way?"

The Centre developed a reputation for being light on its feet, and teachers, visitors and volunteers observed this with delight. "I've always been amazed by how quickly things happen at Fort Whyte," remarked one volunteer. "You'll go to a meeting and somebody will make a suggestion. Then you'll come back the next day and it will already be part of the program."

When the Fort Whyte project started, lobbying for the environment was considered controversial. In 1970, for example, the first Earth Day was modelled after the Vietnam War protests of the time, and was greeted with scepticism by the media, politicians and other members of the establishment. By 1980, Earth Day had grown into a major international event. More than a hundred countries and untold millions of people around the world gathered every April 22nd to show support for the protection of nature. All those kids that Scarth and Muir had put their faith in were growing up, and they were bringing very different new values to the culture. In North America, the "back to the land" movement was in full swing, and magazines like *Harrowsmith* and *Mother Earth News* were thriving.

Lieutenant Governor, Ed Schreyer, addresses a sold-out Rotarian Clubs of Winnipeg Dinner in support of Fort Whyte Centre for Environmental Education

Then a British scientist named James Lovelock published a visionary book entitled *Gaia*, which proposed that the Earth is actually a living entity, a sort of cosmic leviathan which can be damaged by the tiny creatures inhabiting her. This was a novel idea, and it turned the old models upside down. When Copernicus first argued that the earth revolves around the sun, rather than the other way around, he was reviled as a heretic. No one proposed that Lovelock be burned at the stake, but his thesis did invert some old ideas. Throughout history, humanity's fundamental myths had involved themes of "Man versus Nature," with the puny works of man always being overwhelmed and defeated by the power of Nature. For the first time, human beings began to consider the possibility that humanity was now the dominant force. New stories began to emerge, in which humanity was the victor and Nature was the victim. Instead of the vengeful whale that destroys the whaling ship in the 19th century parable *Moby Dick*, the tables were turned in new 20th century parables like *Free Willy*, a tremendously popular Hollywood movie in which a city boy helps a killer whale escape from a zoo.

An original member of the Volunteer corps, Walter Shelmerdine, leads his 'brood' of students to another lasting cattail marsh memory.

In books, movies, and music – in fact throughout western society – there were indications that people were undergoing a fundamental change in their attitudes about Nature. Scientists and environmentalists were beginning to realize that human activity was damaging the global environment, and not just superficially, but to such an extent that humanity itself was in danger.

It's fair to say that the Fort Whyte Centre felt these concerns keenly. And the Trustees spent a lot of time talking about the best way to address those educational challenges. Environmental activism is sometimes seen as a preoccupation of political left-wingers. But the founders of the Wildlife Foundation saw their mission as practical rather than political, and experience demonstrated that support for environmental issues wasn't necessarily going to come from any particular group on the political spectrum.

The issue was education, not political affiliation, and sometimes you couldn't educate people without rocking the boat. Scarth, for example, was sounding the alarm on the dangers of global warming long before it became newsworthy. Even then, (20 years ago) Scarth was insisting on putting out annual reports that included scary-looking maps and graphs illustrating the projected impact of climate change in the prairies. This may have caused some sceptics to wonder if Fort Whyte was a turning into a hotbed of radical environmentalists, but those alarming projections have in fact turned out to be conservative. The causes of global warming may still be under diminishing debate in some shrinking circles, but rapidly receding glaciers and polar ice caps, longer hotter summers, and rising sea levels have turned the hypotheses of the 1980s into the alarming scientific reality of today.

Scarth and other members of the board saw this coming, and believed that Fort Whyte needed to examine its long term goals in light of future environmental issues. "We had proven that we could attract visitors," says Scarth. "Over the last decade, more than 100,000 students had visited the grounds. But it seemed like a good time to re-examine our priorities. The original idea behind the Centre was to provide facilities for education. Conservation was the goal, and the approach was very much a scientific one. But given all the changes that were coming in society, it seemed that we should be moving beyond the role of being a simple 'nature centre' and becoming a broader 'environmental education centre.'"

It was not a subtle distinction. Instead of just strolling in the fresh air and looking at birds and wildlife, it was important that citizens, particularly young citizens, started learning about the social and scientific issues that affected the future of those wild creatures and ecosystems.

In taking that next step, it was important not to move too fast. As Don Muir emphasized in the early days of the project, it was critical for Fort Whyte to avoid the temptation of outgrowing its capacity. About 5,000 kids were visiting the seasonal Nature Centre every year, and the staff and volunteers were struggling to meet the ever-increasing demand for environmental programming. Sam Fabro was handling a lot of the management work. It was a happy arrangement for Fort Whyte but in fairness Sam couldn't be saddled with it permanently.

The property felt small, like an old pair of britches that no longer fits the young owner. Busloads of kids were arriving at the Centre every day, and since the ideal teaching unit for environmental education consists of 10-15 students, it was becoming a challenge to find the staff and the elbow room to provide the kind of high quality experience to which the Centre was committed.

In short, Fort Whyte needed to undergo a major change in both outlook and physical size if it was going to grow to adulthood.

In 1982, Fort Whyte's 10-year lease from Canada Cement had run its course. And thankfully, the arrangement had not proven an embarrassment to John Redfern.

CHAPTER 4

A REALLY BIG DEAL

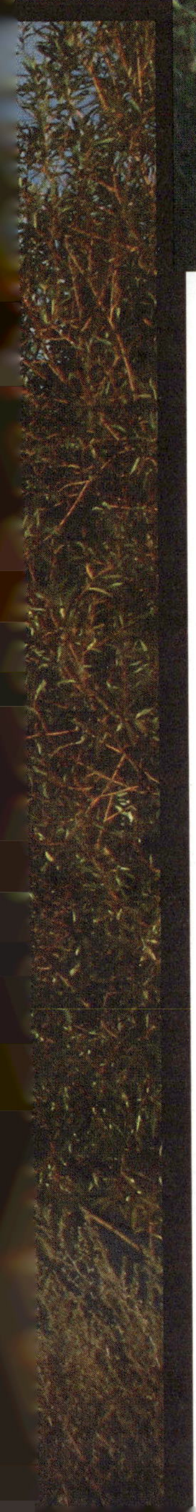

Fort Whyte was still on good terms with Canada Cement, and the trustees did not anticipate any problems in getting the company's support for another step forward. The trustees envisioned a new mission for Fort Whyte and a much larger place, with a theatre for lectures and films, a professional meeting room with a view of the lake, a large freshwater aquarium populated by pike, bass, walleye and other local fish species, and perhaps an indoor waterfowl room that not only housed over-wintering birds but allowed members of the public to watch the ducks preen, socialize and cavort on snowy winter days.

All these ideas required more land and larger facilities. There was an excellent site well north of the Kiwanis Reception Building, on the edge of the relatively immense "Lake Four," with forest on one side and prairie on the other – all in all, 200 acres of fine wildlife habitat. But it was owned by Canada Cement, and the Wildlife Foundation of Manitoba (which still managed the Nature Centre) did not have money to buy it.

If Fort Whyte was going to make a play for that land it had to move quickly, because Canada Cement was considering selling the property to residential developers. Already, the bush land along Wilkes Avenue and Shaftesbury Boulevard was long gone. Wooded bluffs and open trails once populated by deer, foxes and other wildlife had been converted to suburban streets and boulevards. The wave of urbanization was coming ever closer, and if Fort Whyte was to preserve its foothold in the area and preserve what was left of that vanishing wildlife habitat, something had to be done.

To educate himself on the development possibilities, Alan Scarth went to the land titles office and searched ownership of the property in the area. Just to the north of the 200 acres that were owned by Canada Cement he discovered a swath of property owned by the Provincial Crown. The former NDP government under Ed Schreyer had purchased the land with the intent of turning it into low-income housing. But Sterling "Red" Lyon and his Conservatives were now running the show, and Scarth hypothesized that Lyon wasn't interested in being a landlord. Lyon was a Portage la Prairie boy, a hardnosed roll-up-your sleeves conservative, duck hunter, and friend of several Fort Whyte patrons, including City Commissioner D.I. MacDonald, who happened to be a senior member of Fort Whyte's Board of Trustees. As an added incentive, the land in question was probably more valuable for its underlying clay than for its housing potential, and Canada Cement was running out of clay. Scarth added up these coincidences and decided it all sounded promising. Was it possible that there might be potential for a little land swap here?

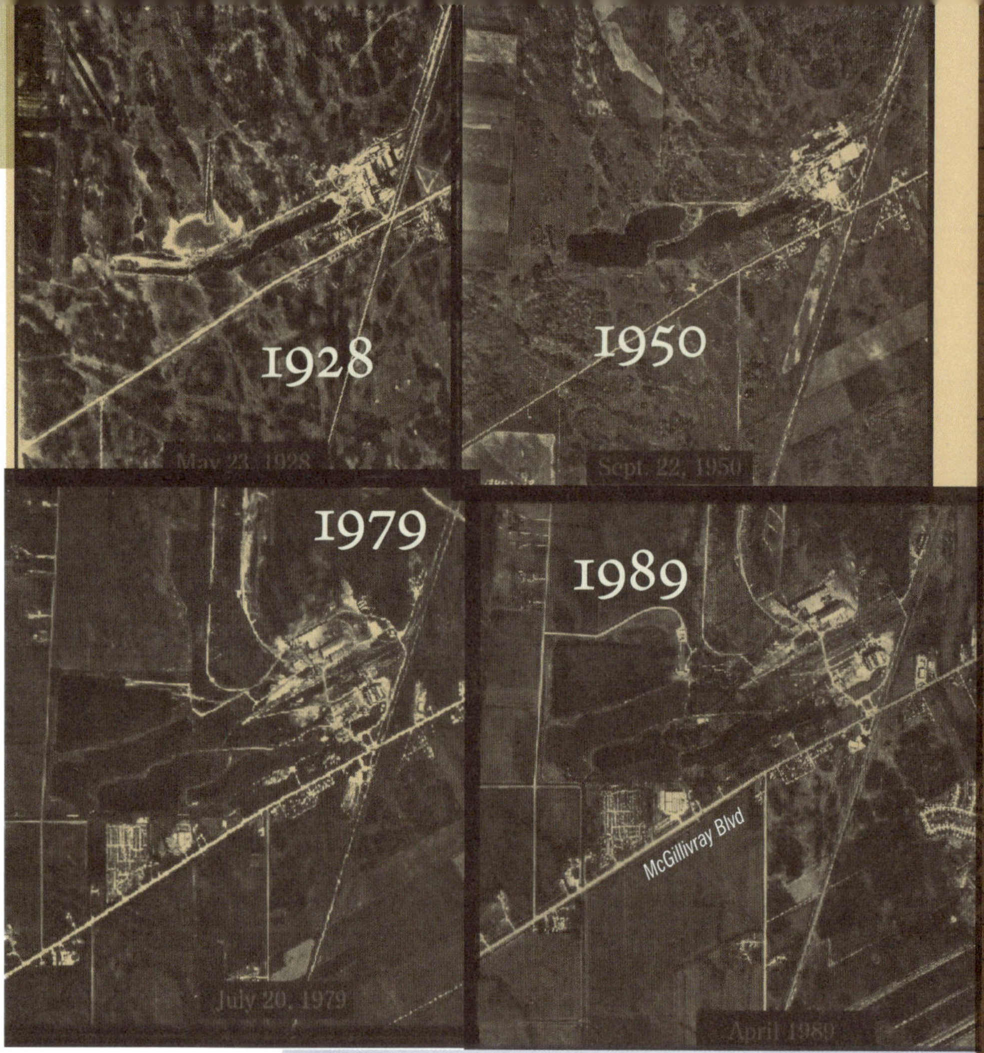

Changing Landscape – From aspen parkland to Winnipeg's 'Lake District', Canada Cement's open pit mining for clay unwittingly carves out Fort Whyte's future home

Scarth picked up the phone and cold-called a young politico named Gary Filmon, who was working for Red Lyon as Minister of the Housing and Renewal Corporation. Scarth asked for a meeting and Filmon agreed. A few days later Scarth showed up at Filmon's office in the Legislative Building with a briefcase full of documentation and building designs. There was a lot riding on the meeting, so Scarth was keyed up. But Filmon had a can-do reputation, and Scarth was more optimistic than he had been ten years ago, getting ready to meet Canada Cement's Harold Edmonds. Upon entering the office, Scarth shook hands with the chipper, athletic-looking Filmon, who despite the hot summer weather was dressed in his standard dark business suit. Filmon and Scarth immediately hit it off. Filmon was widely regarded as an energetic individual with an impressive resume in both business and politics, and Scarth sensed that he was dealing with someone who would rather devise solutions than obstacles.

"Alan was very energetic and persuasive," recalls Gary Filmon. "And I certainly gave him my full attention."

The gist of Scarth's proposal was this – Canada Cement's 200 acres had been mined of its clay and had no more value to the company. The 80 acres owned by Manitoba Housing and Renewal had plenty of clay, and was close enough to Canada Cement that it conceivably could be used as a source of material for the plant. What if Canada Cement swapped its exhausted quarry lands for the 80 acres owned by the MHRC? And what if the Province then leased that acreage to Fort Whyte for use as an environmental education centre?

Filmon thought the idea had some merit. "Alan was very keen on showing me how the project would have important benefits for Manitobans. He brought along impressive architectural drawings of the interpretive centre they hoped to build and it looked very promising. I thought it was a great concept, and asked him what I could do to move things along."

Two decades later as Premier, The Honourable Gary Filmon, continued to join in many Fort Whyte milestone celebrations

Scarth asked Filmon if he could come up with a price for the land. Filmon did some research and determined that the MHRC land was worth about $600,000. Filmon then consulted with his colleagues in government and agreed that the province would be willing to swap this land for the 200 acres of land owned by Canada Cement, and then lease that land for one dollar to the Wildlife Foundation to 'manufacture' environmental stewards and conservation-minded youngsters instead of cement. It looked like a good deal all around. The provincial government would generate some excellent publicity for itself for contributing to the environmental education of the province's youngsters, and Fort Whyte would be sitting pretty with 200 fabulous new acres of lake, prairie and woods.

"We would look good too," adds John Redfern, who was still managing the western region for Canada Cement. "The environmental movement was really beginning to look askance at industries like ours, and this was a chance to get involved as good corporate citizens at no cost to us. Plus we got a whole new source of clay."

Now all that remained was the thorny little detail of the tax bill. As any homeowner knows, it costs money just to own land, even if it is paid off. If the Province leased the land to the Wildlife Foundation for a dollar, the Foundation would still have to pay a considerable amount every year in municipal taxes. The Foundation had not yet developed the $2 million in cash flow it enjoys today, so that tax liability would have scuttled the deal if Gary Filmon hadn't come up with a solution. He suggested that the Province would give the City of Winnipeg a transfer grant which would get Fort Whyte off the hook for the municipal taxes. This again would have certain benefits to both levels of government.

Overall, it was a brilliant concept, one that truly embodied "the art of the deal." But as usual, the devil was in the details. "We had an agreement in principle," says Alan Scarth. "But it took months to sort out issues like the boundary between the plant property and the 200 acre piece, the preservation of water supply to the cement plant, access roads, zoning, and myriad other concerns."

A senior official named Bob Clarkson worked on the deal from the provincial government side and D.A. Thompson, a veteran lawyer with the old Winnipeg firm Thompson Dorfman Sweatman burned the midnight oil on the Canada Cement side. "You really can't say enough about how these two men leaped hurdles and cleared an absolute wilderness of legal underbrush to make the deal happen," says Alan Scarth. "Clarkson was an excellent administrator who was genuinely interested in seeing a good transaction move ahead. Because the swap was so complicated it took many months to conclude, and in fact the final deal wasn't consummated until after the Conservatives under Sterling Lyon had been defeated and replaced by the NDP, so Clarkson's steady hand and patronage of the project was essential. All of which demonstrated how Fort Whyte was never one person's vision. It's always been a great collaborative effort. If it wasn't for the tireless work of hundreds of individuals there are many times the project would have stalled and we wouldn't be where we are today."

Rough earthworks to fine landscape...giving Mother Nature a headstart on reclamation

Finally, the deal was all wrapped up and it seemed time to celebrate. Scarth and the rest of the Fort Whyte team got to work planning a ceremony to celebrate the acquisition of the new property. Then the phone rang. It was Clarkson. "He said we had run into a glitch," says Scarth. "It was a grand understatement, and typical of this unflappable fellow."

In the process of preparing the transfer of the 80 acres to Canada Cement, Clarkson had discovered that the Province of Manitoba didn't actually own the land. Not all of it, anyway. The province just owned the "surface rights."

It's a little known fact of law that mining rights hold precedence over most other property rights. A mining company can show up and set up a mining operation in someone's front yard if they happen to own the mineral rights. In this case the rights to the minerals below the surface were owned by persons unknown.

Canada Cement's only reason for agreeing to the swap was to acquire a new source of clay. Without the mineral rights they would no doubt cancel the deal and go back to their first plan, i.e., to sell their 200 acres of mined-out clay pits to residential real estate developers. And no one could blame them. Canada Cement was after all a business.

Once again, for the umpteenth time, it looked like Fort Whyte was going to lose the game in the final inning. But Scarth wasn't ready to quit. He researched the archives and determined that Winnipeg businessman who owned 50% of the mineral rights. The other 50% were held by a woman by the name of Slater, whose address, telephone and whereabouts were not listed.

Scarth says he didn't see any point in "bothering his friends at Canada Cement with the news about the glitch." Instead, he allowed the plans for the celebration ceremony to go ahead. Then he privately set off on an emergency campaign to acquire the mineral rights.

"I called the Winnipeg businessman and he was amenable to a deal. I suggested we might value his 50% interest at $12,500, and that he might consider the possibility of donating it and receiving a charitable receipt. The figure was conservative to say the least, my thought being that if he was going to donate his interest, he would not care that much about the valuation of it, and the modest valuation would be useful as a basis for negotiation with Mrs. Slater, if and when she could be found."

Scarth sleuthed out the identity of the mysterious Mrs. Slater and determined that she was living in Lethbridge, Alberta. He jumped on a plane to Calgary, rented a car and drove to Lethbridge. By telephone she had agreed to meet with him and he was hoping to close a deal on the spot. At the front door of her apartment he was greeted by a tall dignified woman who invited him inside. She told him she had once lived on a farm in Manitoba, and she was quite familiar with the land in question. To all appearances she seemed like an individual who could use some cash. She lived alone, and Scarth noticed that she kept a phone on the floor so it would be accessible if she fainted. "I made my usual pitch," recalls Scarth. "I asked her to join us in educating school students about keeping the environment healthy, but she was flinty-eyed as she listened. When I was finished she said. 'Son, we each have our charities. You have yours, and mine is the Christian message.'"

She went on to describe her efforts to support an evangelical radio station in Toronto. "I couldn't argue with this, so I was reduced to playing the money card, which wasn't exactly an ace. I offered her $12,500 for her interest, saying that a local business man had accepted the valuation, but she said, 'That's chicken feed, and you know it.'"

Marsh Marigold

NATURE NOTE

The Marsh Marigold, Caltha palustris, is a beautiful springtime sight. Like many beauties, the Marsh Marigold harbours a dangerous secret – the foliage, roots, and flowers of the plant contain poisons that can bring on heart attacks, and the plant's sap can blister skin and irritate eyes.

Sterling example of sustainable development – view from Fort Whyte's new wind turbine looking southeast over the Canada Cement plant with the Waterfowl Gardens landscape in the foreground.

Scarth knew indeed that it was chicken feed, but chicken feed was all he had. She ushered him to the door, wishing him well with his campaign, and he wondered if this was the end of Fort Whyte's dream of future expansion. Although he was disappointed, he couldn't help but admire the woman's fortitude. "My lasting memory of Mrs. Slater is shaking the hand of that tall, valiant woman at the door of her small apartment where she was living out her life, a long way from her Manitoba farm."

He went back to Winnipeg and was trying to figure out how to raise more chicken feed when we heard the news that Mrs. Slater had died. "I never did hear the details of her passing," he says, "But perhaps the poor woman fell and could not make it across the floor to the phone."

He did some follow-up research and found out she had left the mineral rights to the Toronto evangelical radio station. Scarth called the station owner and offered him $12,500 for the rights to the clay. "For him it was money for jam, as the English say, and we had a deal."

Scarth no sooner tied up this side of the deal than the other half started to unravel. "It seems that our Winnipeg businessman had run into a problem with a ski resort investment in Aspen, Colorado, and he was no longer interested in donating his share of the mineral rights to the Foundation. But he was still willing to abide by the original price, so we emptied out the cookie jars and came up with a payment of $12,500."

The signing ceremony went ahead as planned, right on schedule, with the officials of Canada Cement and the Province of Manitoba blissfully unaware of the fact that the Province was trading land to Canada Cement that would have been utterly useless if Scarth hadn't negotiated the mineral rights in the background. Nor did Scarth ruin the occasion by revealing that the Provincial officials had come within a hair's breadth of embarrassing themselves in front of the media by trading away property rights they didn't own. "It was a small matter," says Scarth. "And we didn't want to trouble anyone with it."

Minister of Natural Resources, Leonard Harapiak, seals the deal with the signing of the 100-year land lease to the Wildlife Foundation of Manitoba. Alan Scarth's tireless negotiation efforts are finally rewarded amidst a theatre full of daffodils on January 29, 1987.

It would have been understandable at this point to assume the negotiation was finally resolved. But like a suspense story with multiple endings, each false bottom of this deal was no sooner dealt with than it revealed another. The provincial land package, as the reader may recall, was originally acquired by Ed Schreyer's NDP government as a potential site for low-income housing. When Lyon took over as Premier of the Province in 1977, he immediately applied many of the free-market principles that were being introduced by Margaret Thatcher in Britain. In 1981, the political pendulum in Manitoba swung back towards the left, and Lyon's government was toppled by Howard Pawley and the NDP. The Province's new minister of Natural Resources, The Honourable Sam Uskiw, needed to approve the lease before it could be finalized. The land swap was one of those little surprises that government officials often find on their desks after the last administration has left the building.

Uskiw admired the work of Fort Whyte, but in the words of Alan Scarth, "He was also very shrewd and he understood deals. From his perspective, the Tories had traded valuable MHRC residential real estate for 200 acres of spent quarry land which was then being given to the Fort Whyte Foundation for a mere dollar, and free of municipal tax to boot."

Uskiw might reasonably have argued that the previous administration had literally sold the family farm for a dollar, and his simple refusal to sign would have obliged the Wildlife Foundation to come up with some money to sweeten the deal. But Scarth says Uskiw was a good guy and wasn't about to ruin the celebration by haggling over a tax-free gift of $600,000.

"The deal was so far along that Sam Uskiw agreed to sign the lease," says Scarth. "It was a great relief. Finally, the 200 acres of prime wildlife habitat was ours."

CHAPTER 5
MINING FRIENDS & SUPPORT

The next stage was to convert the grandiose architectural drawings into reality.

Scarth had managed to acquire the new acreage for a dollar. But it was unlikely that similar feats of horse-trading and sleight of hand would work this time. The new Fort Whyte Centre would require money, and lots of it.

Eric Harvie, who became the West's most generous philanthropist, at the Meridian No. 1 well near Ribstone, Alberta, 1930

As it turned out, the cash was buried underground – about a mile underground – under a patch of scruffy Alberta rangeland 800 miles to the west. The money had been buried there for about 100 million years. During the Mesozoic era, when dinosaurs like *Stegosaurus* and *Brontosaurus* walked the earth, the dominant life form on Earth was (and still is) tiny, single-celled organisms called diatoms.

Diatoms are the bottom rung of the food chain for all marine and freshwater animals. When studied under a microscope, they look like jewels. They can be triangular, fan-shaped, or twisted like curls of pasta. In a few acres of water, there are billions of diatoms. They have short life spans, and when they die their carcasses drift to the bottom and form a kind of silt. Over millions of years, these layers of diatomaceous silt can become buried and compressed by the tremendous weight of the earth above. We're all familiar with the methane smell of a healthy marsh – a smell that hints of the gaseous potential of decomposing organic matter. Deep pockets of ancient diatomaceous earth were trapped under the Alberta rangelands, and over many millions of years, intense heat and pressure slowly transformed them into natural gas and oil.

In the 1940s, Imperial Oil was trying to find oil in Alberta but wasn't having much luck. Canada was buying most of its oil from the United States and Calgary was a small town scratching out a living from the livestock business. During the first half of the century, Imperial Oil and other companies drilled more than 1,000 wells in Alberta at a cost of around $200 million. But the oil companies had little to show for their investment. In 1946, the company's directors met in Toronto and faced the facts – the company had drilled 133 "post holes" without drawing a drop. It was obvious that Alberta contained no oil reserves of consequence and in the opinion of the company directors it was time to move on.

During that grim meeting, a few senior members of the exploration team pleaded with the directors to be allowed to finish the drilling season. There was a major argument in the board room, but the directors reluctantly agreed to pay for one more hole before scrapping the Alberta program. Head geologist Ted Link went back to Alberta and took a poll of his 32 scientists. By consensus, they decided to drill their last hole near the town of Leduc. Their tests had detected a bulge of Cretaceous limestone far underground that might be worth checking out. The odds were slim, but they figured that since they had the budget to drill one hole, they might as well drill into an anomaly.

The designated tool pusher was a colourful character named Vern "Dry Hole" Hunter, whose main claim to fame was jinxing every drill he ever touched. All the way down to 2,000 feet, Dry Hole and his crew picked up little signs that this well might be different from the others. They kept detecting vague rumblings and smells. "We started to pick up signs of gas and oil in the Cretaceous," recalled Dry Hole. "The whole thing felt alive."

When they reached D2 – the Devonian Layer – the well blew, showering the drillers with light sweet crude. The discovery was not only a jackpot for Imperial Oil but it revealed a vital secret. Before then, most geologists believed that oil in central Alberta would be found in the shallow Cretaceous layer. But the jackpot in Leduc proved that most of the oil actually resided in the Devonian layer. The discovery led to the development of a series of major oil fields in Alberta. Over the next decade, Alberta's oil output rose from 6.7 million barrels annually to 144 million, and Canada was well on its way to asserting its independence and influence by becoming a major international oil producer.

During that same period, a Calgary lawyer named Eric Harvie was acting for various exploration companies, and had been smart enough at the early stages to waive part of his legal fees in lieu of royalties on whatever the companies discovered. When the boom hit, Eric Harvie made $50 million. Towards the end of his life, he decided that he wanted the money to go to a good cause, so he established a funding body called The Devonian Foundation. Most charitable trusts establish a nest egg and use the annual interest to support worthwhile projects. But in an unusual move, Harvie decreed that the entire capital of the Foundation was to be paid out in western Canada by the 1980s - right around the time that the Wildlife Foundation of Manitoba was trying to find someone to support the new expansion at Fort Whyte.

Wild Bergamot

NATURE NOTE

Though the honour of flavouring Earl Grey Tea belongs to an Asian citrus fruit with a similar name, Wild Bergamot, Monarda fistulosa, can be used to make a delicious herbal tea. The pla[nt] grows in moist woods and prairies. It also thri[ves] alongside many of Manitoba's roads, as far north as Norway House.

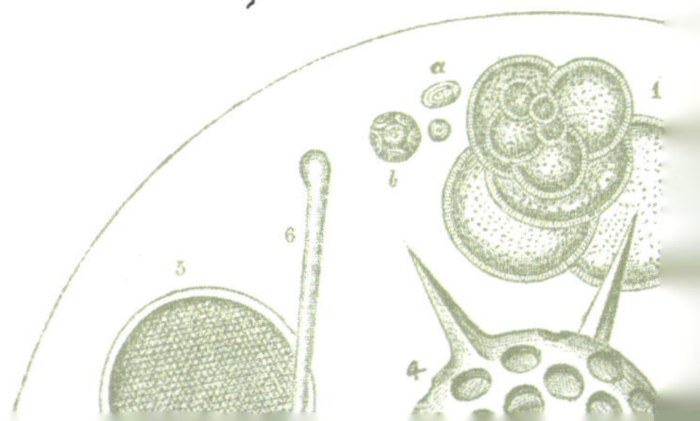

Don Harvie, Eric's son, was managing The Devonian Foundation after his father's death, and he hired two well-intentioned executives, Todd Meech and Jim Fish to act as a granting jury. After getting wind of the Devonian Foundation, Sam Fabro and Alan Scarth flew to Calgary and met with Jim Fish and Todd Meech in the office of Devonian. "Meech was a visionary," says Scarth. "And he was intrigued with the potential of Fort Whyte. When he and Jim Fish came to Winnipeg to check out our newly-acquired 200 acres, his immediate reaction was that we should buy the acreage around it as a buffer against the encroaching city."

Having just acquired the 200 acres from Canada Cement, the Fort Whyte trustees were feeling land-rich and cash-poor, and accustomed to moving forward methodically, they didn't want to get ahead of themselves. But Fish's advice proved solid. "Years later, we were finally able to follow his counsel," says Scarth. "But by that time we had to pay $3,000 an acre."

Jim Fish handled the business side. He was responsible for laying down the conditions for the grants from Devonian, and when Scarth asked him for a nice round figure of a million dollars, Fish didn't blink.

"He agreed to the figure," says Scarth. "But he stipulated that Devonian was only prepared to give us the money if we could match it with contributions from our Manitoba friends. Over the next few months, we scrambled to make up the shortfall, and that was good discipline for us."

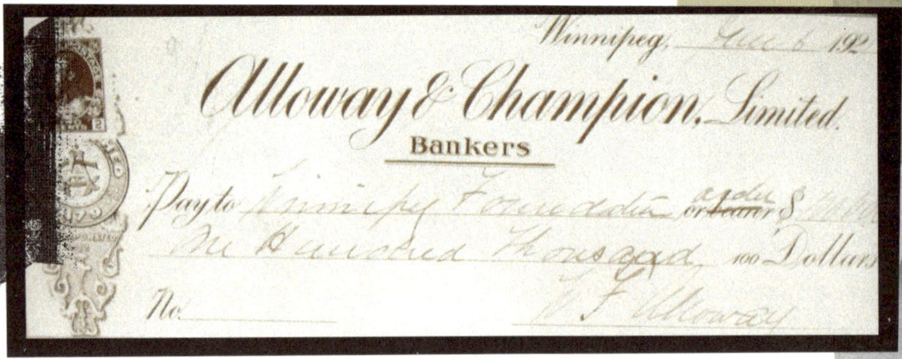

William Forbes Alloway

Like any crisis, the fund-raising blitz helped to clarify the identity of Fort Whyte's true friends. George Richardson was headed overseas when he heard about the funding opportunity with the Devonian Foundation. Just before he boarded his aircraft for Europe, he called Alan Scarth from the airport and said that Fort Whyte could count on him for $150,000.

"Keep in mind that his was a quarter of a century ago," says Scarth, "when $150,000 was a very considerable sum, and the million dollars we needed in order to match the Devonian grant was an almost impossible sum to raise in a city the size of Winnipeg."

Winnipeg may be small, but the city has a long tradition of charitable giving, and without it the Fort Whyte Centre could never have raised the matching funds. As always, The Winnipeg Foundation showed itself to be a steadfast friend. Established in 1921, with a gift of $100,000 by William Forbes Alloway, The Winnipeg Foundation was the first community charity in Canada. The idea was that individuals would donate money to the Foundation and the interest generated by the pool of cash would be used to support worthwhile charities. Over the years, thousands of individuals have given money to The Winnipeg Foundation. Some of those gifts have been modest (in 1924, a gift from an anonymous widow of three gold coins worth a total of $15 became a celebrated endowment, symbolizing that "no gift is too small.") And some gifts have been breath-taking. (Like the $100 million handed over by Randy Moffat – the largest gift ever made to a community foundation in Canada.) But they all added up to make The Winnipeg Foundation the second largest community foundation in Canada – an extraordinary fact given that Winnipeggers are nowhere near the top of the list in terms of average per capita income.

In 1980, the constitution of the Winnipeg Foundation was revised so that its grants could go out to organizations working to conserve human, natural and heritage resources, and this revision occurred just in time to allow the Foundation to support the drive to match the Devonian funds.

On September 27, 1983, at the opening of Fort Whyte Centre, H.D. (Des) Whellams passed on his torch – the key to the Centre door – to four-year old Billy Elliott, representing the youth of Manitoba to whom the Centre was dedicated

Green-winged Teal

As Alan Scarth puts it, "The Winnipeg Foundation was a steady ally of Fort Whyte's thanks to the leadership of its Executive Director, Alan Howison. He caught the Fort Whyte vision very early in the process, helping out with moving and retrofitting the old Waterfowl Wintering Building and other crucial expansions. Nowadays Fort Whyte might attract 100,000 visitors a year, including 30,000 school students, so from today's perspective it seems like an obvious project for The Winnipeg Foundation to support. But 25 years ago, the word 'environment' was just coming into use. And the idea that we were trying to put together a Centre that might attract so many people was considered by some to be ludicrous, and a recipe for insolvency."

Scarth says that in the early 1980s, only a limited number of people saw the potential importance and mass public appeal of environmental education, and it took both vision and courage to support that vision with a cheque. "Fortunately, Alan Howison was not short on courage," says Scarth, "and it was very lucky for us that he was in our corner."

NATURE NOTE

The Green-winged Teal, Anas crecca, is more at home on mudflats and shallow ponds with emerging vegetation than on more expansive marshes. This teal is a circumpolar species, with some Eurasian Green-winged Teal showing up in North America every year, and vice-versa.

It might have seemed impossible to raise that money in a city the size of Winnipeg, but raise it they did. The Carolyn Sifton Foundation, Harlequin Enterprises, Imperial Oil, Ducks Unlimited, and aforementioned donors like George Richardson and the Winnipeg Foundation all did their part. At the end of the fund raising blitz, the Fort Whyte trustees were able to bring a million dollars to the table to match the funds offered by the Devonian Foundation. The architects were called in, and plans were launched to build a fabulous new centre.

To recognize its new direction as a champion of the environmental movement, the new complex would be called the "Fort Whyte Centre for Environmental Education." It would have the quintessential Manitoba setting – aspen forest on one quadrant, marshland to the south, open prairie to the north, and the open, wind-whipped waters of Lake Four to the west.

To honour Eric Harvie, Dry Hole Hunter and the diatoms connected to that prehistoric oil deposit under the Alberta rangeland, Lake Four would be renamed Lake Devonian.

(Left) Since seizing the Fort Whyte vision as a Founding Trustee in 1966, George Richardson has volunteered valued counsel and encouragement through four decades of dynamic growth in programming, facilities, and land.

(Right) The Winnipeg Wood Duck Project and its pioneer nestbox custodian, Elliot Rodger, introduced Kay Lightcap to Fort Whyte, and her relationship with the colourful little cavity-nesting duck eventually led to her making a major contribution to the Aquarium of the Prairies capital campaign. She subsequently established a philanthropic fund with the Winnipeg Foundation focusing on education, and the advancement of young people.

Upper Mandible
Nostril
Forehead
Crown
Back of Head
Back of Neck
Back
Secondaries
Coverts
Belly
Tibia
Tarsus
Inner
Middle
Outer
Primaries

Birds

BIRDS
AVES
PASSERES
Finches
Starlings
Crows Jays Magpies
Larks
Tyrant Flycatchers
Hummingbirds
Swifts
Goatsuckers
Woodpeckers
Kingfishers
Trogons

CHAPTER 6

DESIGNING WITH NATURE

The trustees knew there was going to be a new centre, but where would it go? And what would it look like? The field of building design had been evolving just as quickly as the field of environmental education, and one of the most forward-thinking designers in Manitoba was a young landscape architect named Garry Hilderman.

Hilderman was raised in Winnipeg, the son of an electrician, and his dad, like most fathers, wanted him to specialize in a respectable, solid field that would provide him with a good living. "He wanted me to become an electrical engineer," says Hilderman, "which didn't appeal to me at all. I wanted to be a fine artist, which appealed to him even less. So we split the difference and I went into architecture."

Architect's model of the proposed Fort Whyte Centre

As a teenager, Hilderman had summer jobs working for a landscape construction company, which seemed to satisfy his love of plants and dirt. Hilderman studied architecture at the University of Manitoba, but while in the fourth year of a five year program, the architecture students were exposed to a landscape architect from England who taught a course in site planning and suddenly he knew what he wanted to do. So, he took his credits and enrolled in the Landscape Architecture program at the University of Michigan.

After working as a landscape architect for firms in Toronto and Winnipeg, he went to graduate school at the Berkeley campus of the University of California during the tumultuous days of the 1960s, and was intrigued by new ideas about the relationship between the "built" environment and the natural environment. "It seems a bit abstract," he says. "But it really isn't. If you look at the average downtown office building, it's the same on all four sides. But the northern side faces a very different world of nature than the southern side. The sun rotates across the southern face and the coldest winds strike the northern face. So just on the practical level we should be thinking about this. Architects talk about form following function, but form also effects function either for good or bad.

The concepts might seem obvious today but North American architecture in the early 20th century was heavily influenced by the Victorian era, which ignored the natural setting and liked to devise architectural symbols of "Man's Dominion over Nature." However, architects like Frank Lloyd Wright were talking about and practicing "organic architecture" and landscape architect Ian McHarg was writing about "Designing with Nature." With the 1960s came the birth of the environmental movement and Hilderman was intrigued by "the challenge of designing buildings and sites that related in a practical and spiritual way to their natural settings".

While studying at Berkley, Hilderman convinced Walter Danyluk, the first superintendent of Manitoba Parks to give him a summer job studying the landscape of the relatively new Whiteshell Provincial Park using techniques he was learning at Berkeley. He spent the summer of 1968 undertaking a resource analysis of the Park. "They gave me the use of Beaver floatplane and a pilot and we spent the summer flying the shorelines of each lake and recording their natural features and I interviewed every resource expert in Manitoba who would talk to me from entomologists to archaeologists".

That summer work became the foundation for his Masters thesis, which described a rationale for planning and developing for recreation in the boreal forest - the interface between man and nature. Hilderman describes that summer with pleasure. "After flying shorelines all morning we would look for a party of fishermen with their guide cooking walleye over an open fire. We would land and they would feed us a great meal. After lunch and conversation we would climb back into the Beaver and fly off. It was a dream summer job."

After graduating from Berkeley, Hilderman returne to Winnipeg and started his own practice of Landscape Architecture and Planning. He also taught part time in a brand new Department of Landscape Architecture in the Faculty of Architecture at the University of Manitoba. There he met Alan Scarth when they sat as advisors in a student practicum committee. Scarth had an idea for an expanded environmental education facility at the Fort Whyte lakes and needed a feasibility study. He asked Hilderman to help him with the task. "We were both interested in the same issues. The study was entitled 'A Feasibility Study - The Fort Whyte Centre for Environmental Education' and formed the bases for the first phase development of what eventuall evolved into today's FortWhyte Alive."

Hilderman's study of Fort Whyte inventoried the features of the existing property and identified a rudimentary program for environmental education that could take advantage of the site and Scarth's ideas. They laid out a plan for a new visitor centre that

would do justice to the concept for environmental education. Ideally, from the moment a visitor passed through the gates on McCreary Road, every physical element of the property would in some way manifest a commitment to the environment. As Hilderman explains it, "I argued that the new centre building be set back from McCreary Road, well away from the entrance. There needed to be a transition away from the city to an experience of natural forces. A curved entrance road would create a sense of departure from the geometry of the city. All aspects of development of Fort Whyte would reflect an understanding and respect for the natural world."

The visitor centre building was conceived to face south with a prominent overhang which would provide shade from the high summer sun, but would also allow low-hanging winter sun to enter the building. The roof was to slope down to the north to deflect winter winds and the materials used would reflect the conservation ethic – woods like cedar, that didn't need preservatives, and would weather differently due to orientation.

Hilderman believed that the building should afford visitors a variety of perspectives of the landscape – from vantage points above the land to ground level. The new building should therefore step into the ground providing a variety of views and as well give a sense of protection from the harsh prairie climate. After Hilderman completed the Feasibility Study, Alan Scarth took the concept and colour renderings for a new environmental education centre on the road looking for funds to build the vision.

Once the project had enough money to proceed, the work of preparing detailed design and construction drawings began. Scarth oversaw every detail of the process. It was his dream and he knew when it was being realized and when it wasn't. Carl Nelson, an architect teaching in the Department of Landscape Architecture at the University of Manitoba, was asked to turn the general concept for the building into a specific and inspiring form. Carl Nelson's role in shepherding the project through the many phases of design and redesign cannot be overestimated. He and Hilderman volunteered hundreds of hours of their time to ensure that the buildings and the grounds expressed the ideals of environmental stewardship. As Alan Scarth puts it, "Carl Nelson was a crucial player throughout this long involved process. I'm not overstating the matter to say that we could not have made the project happen without him. His design genius and supervision of construction produced a spectacular building, for which he deservedly received the prestigious Premier's Design Award."

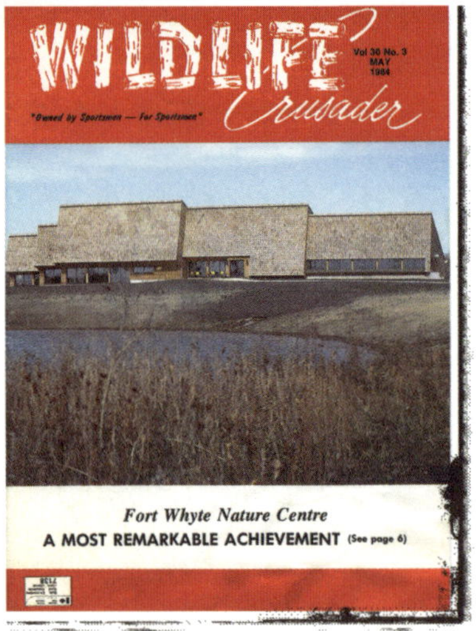

MMP Architects were commissioned to prepare detailed construction drawings and specifications, and Malcom Construction was hired to build it. While the building was under construction, the lakeshores of Fort Whyte were being regraded to make them less steep and safer, and a water control system was being developed to prevent flooding. The building was stepping down into the land, but hopefully not so far down that seepage and even flooding would be a problem. Local farmers were telling of years when all the farm lands around Fort Whyte were inundated in the spring, Hilderman had many a sleepless night as he went over the geodetic calculations in his mind, hoping the site and the building would be safe from spring flooding.

There's a long-standing tension between architects and builders. Architects are not always skilled in the fundamentals of construction, and builders are not always patient with the theories of architecture. A builder tends to believe that the main task at hand is to put up a solid, weatherproof building that functions efficiently and doesn't come in too far over budget. An architect often sees a building project as an opportunity to convey ideas or send a message. As the great Chicago architect Henry Louis Sullivan famously declared, "Every building is an act," – a lofty notion which no doubt would have riled the legendary movie producer Sam Goldwyn, who once remarked, "If you want to send a message call Western Union." In a difficult project, both parties tend to be impatient with the other's role. The builder tends to think that it doesn't take much know-how to draw a structure on a piece of paper – the real challenge is to build it. The architect on the other hand tends to believe that anyone can nail together a weather-tight box – the real challenge is in installing a building that dramatizes interesting and provocative ideas.

The new building at Fort Whyte was going to be an unusual one, and could have become a venue for this age-old antagonism. But in this case, the architect

Interpretive Centre

Carl Nelson and the builder Bud Malcom rose to the occasion. Alan Scarth describes him as "The perfect man for the job. We didn't have to worry about petty squabbling because both sides were represented by first rate people. We all wanted to create something that expressed the conservation ideals of the new Fort Whyte Centre. And Bud Malcom, a quiet, affable man, was dedicated to this unusual building and showed unlimited patience in dealing with the never-ending stream of suggestions from the Board."

By spring, 1983, the new building was rising from a wasteland of lumber, raw mud and excavating equipment on the east end of Lake Devonian. It was an impressive sight, a slope-roofed edifice that would embody the dreams and ideas of the new Fort Whyte. Along with the new building there would be a whole new direction. It was time to get serious. This was a multi-million dollar asset, and it would need to be operated by full time staff and a highly qualified boss. The founders – unpaid supporters like Don Muir, Alan Scarth and Sam Fabro – were ready to step back from Fort Whyte and hand the keys over to a new president. But first they had to find him.

The new job called for a mixed bag of skills. The candidate would need to be a wildlife enthusiast who didn't mind spending most of the time working inside a cramped office. He or she would have to be equally comfortable slogging though the marsh in a pair of chest waders or working the crowd at a stylish social function. The boss would have to understand the founding vision. But it would also be his or her responsibility to steer the Centre in a new direction. It wasn't going to be easy finding the right person. But they had built it, so maybe he would come.

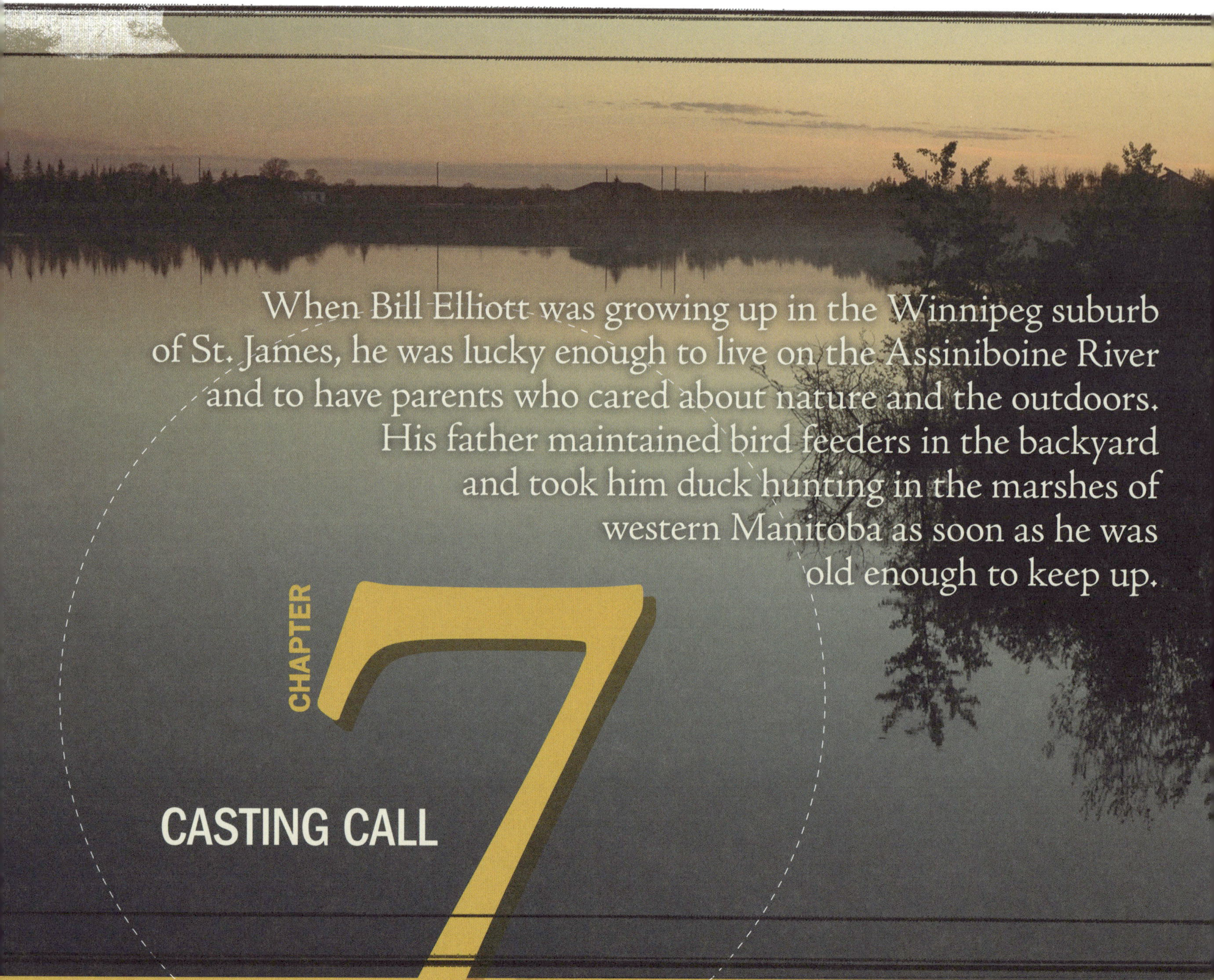

When Bill Elliott was growing up in the Winnipeg suburb of St. James, he was lucky enough to live on the Assiniboine River and to have parents who cared about nature and the outdoors. His father maintained bird feeders in the backyard and took him duck hunting in the marshes of western Manitoba as soon as he was old enough to keep up.

CHAPTER 7

CASTING CALL

Like many youngsters who are introduced to Nature at an early age, he was astounded by the pure exhilaration of being at large in the outdoors. He especially loved duck hunting – the feral rustling of the wind in the high cane grass, the startling brightness of the Milky Way as he and his father waded out into the dark marsh an hour before dawn, and the whistling of unseen wings as the first squadrons of bluebills and mallards swept overhead. “I fell in love with waterfowl and marshes,” he says. “I couldn’t get enough of the sights and sounds of the birds zooming overhead, the marsh wrens checking us out, or the muskrats protesting our trespass. It was my parents’ gift to me and it changed my life forever.”

Bill Elliott, 1990

Back home in St. James he started feeding mallards that staged on the river every autumn beneath their home. He would trickle a trail of grain from the river up towards the house and the ducks would follow the trail right up to their swimming pool. “We had one of the first backyard swimming pools in Winnipeg,” says Elliott. “And the builders would have been surprised to see a hundred and fifty ducks lounging in and around our pool. When winter came, there were always a few hardy ducks that toughed it out and stayed on for the winter. I don’t know where they found water, maybe by a storm sewer outlet on the river somewhere. But every morning, no matter how cold it was, about a dozen mallards would zoom down the river and land in the snow of our backyard and wolf down their breakfast of grain. When they weren’t eating they would stuff their bills under their wings, to breathe warm air, doing everything they could to conserve energy on those bitterly cold mornings.”

In university, Elliott majored in zoology. The wild mallards frequenting his riverbank were handy subjects for his assignments in his Animal Behaviour class. Again, his parents were strong supporters. Allowing Bill to sleep a bit longer, his mother would faithfully record the correlation between sunrise and feeding flight data in January, standing in her bathrobe at her bedroom window in the cold and dark, waiting and watching.

For summer jobs, he did waterfowl and habitat surveys for Ducks Unlimited and assisted graduate students at the Delta Waterfowl Research Station. Upon graduation with a Masters of Natural Resource Management with a practicum on the history and law surrounding wetland drainage in Manitoba, he got a full time job with DU. "I was very fortunate to walk out of University and directly into a newly created position of Habitat Development Biologist," he says.

Within a year he was recruited by the Delta Waterfowl Research Station as their Development Coordinator… a great opportunity to experience the business side of wildlife conservation. "I got to travel a fair bit through the U.S. and Canada sharing Delta's major contribution to the understanding of waterfowl and wetlands with wealthy duck hunters and companies and trying to persuade them to donate money. I lived with my wife and kids in Portage la Prairie and worked at Delta. It was an unbelievable setting and experience rubbing shoulders with the world's leading waterfowl scientists. I got my first hunting dog, spent a lot of time outdoors, and worked with some wonderful people whom I still enjoy socially and professionally."

He had been working and living in Portage for five years when, one spring day in 1983, he flipped open the *Careers* section of *The Winnipeg Free Press* and noticed a job advertisement for someone to run an outfit called the "Fort Whyte Centre for Environmental Education." He wasn't familiar with the Fort Whyte Centre but was faintly aware of Alan Scarth's involvement…a big plus in his mind. And he was curious. Turning to his wife he said, "Read this, it sounds like me."

Coneflower

He sent a resume to Garry Hilderman's office in Winnipeg and did a little homework on his own. With his dad and toddler son, he drove out to Fort Whyte to have a look at the place. "I had no idea that kind of facility existed in Winnipeg," he says. "It was well screened by all a strip of aspen along McGillivray and I was really shocked when I saw the lakes and this big expanse of meadow and woods – very impressive. It was a warm spring day and we pushed the stroller down the mud trail, getting all covered with muck in the process, and had a look at this striking building that Malcom Construction was finishing up. It was a weekend, and I remember being so impressed by the building design and the fact that the crew was hard at work. And they were really nice to us, too, very hospitable, like Bud Malcom as it turned out, which surprised me because we had just walked in off the road…trespassing, actually. Overall, it gave me a favourable impression of what was going on there, and I was doubly excited when Garry Hilderman called me from his office to schedule an interview."

When Elliott showed up, the new Centre was still raw and unfinished on the inside. The Selection Committee, consisting of D.I. MacDonald, Alan Scarth, Jennifer Shay, Boris Hryhorczuk and Garry Hilderman sat at a desk improvised from two sheets of plywood and some sawhorses. "The head of the Selection Committee was D.I. MacDonald, and Alan tactfully whispered that Mr. MacDonald was hard of hearing and I would have to speak up. So I sat there shouting my answers which I could hear echoing off this great cavernous ceiling. It could have been an awkward interview, but they did a good job of putting me at ease, and I felt that I had pretty strong credentials for the job. At that time, waterfowl were the marquee element of Fort Whyte's future, and my whole life had revolved around working with and for waterfowl and wetlands."

D.I. MacDonald

Elliott had other attributes as well. Waterfowl scientists were not hard to find, but most were of a technical bent – researchers who preferred slogging around in the mud all day in the company of birds and muskrats to sitting through meetings in conference rooms or schmoozing potential supporters. Elliott was comfortable in both worlds, and the members of the selection committee were particularly impressed with his easygoing confidence, his gregarious sense of humour, and his solid track record as a well-travelled fund raiser for the Delta Waterfowl Research Station.

There were many applicants, but it was Elliott who eventually got the phone call from D.I. MacDonald, offering him the job. "It felt like a tremendous gift," says Elliott. "I was so young and they were handing me the keys to this wonderful place…and their vision."

His facility with the business world was going to be crucial, because Fort Whyte has always relied on private rather than government money. And in its new life Fort Whyte would require sustained support and lots of it. From his years on the road with Delta, Elliott says he had "a pretty good sense of what it takes to get people to invest in an inspiring vision." He knew the business community is attracted to professional management and measurable results. He knew Fort Whyte was going to have to evolve, and meet the business community halfway. It would have to shed its "pure naturalist" image and get more involved in programming that got average citizens in the door. The Centre needed to become much more than just a waterfowl refuge. It needed to become an active part of the social, cultural, and educational life of the city. Fortunately, the directors and trustees were in strong agreement with this point of view, and they were quite willing to give him the authority he needed to make that happen.

"Alan Scarth was great," says Elliott. "He made it very clear that I was in charge. He literally said to me, 'Take it away. You're the boss.' It was the first time in my life I had been a boss, so it was a little intimidating at first. But overall, the raw potential of the place was so fantastic that I felt like I was signing on for the biggest adventure of my life."

Elliott's first step was to move with his family into the on-site living quarters in the Interpretive Centre. "That was a really important move," he says. "It was not only important as a way of psychologically taking ownership of the place, but it also gave me intimate knowledge of the property. I walked every foot of the grounds and got to know every creak and squeak in the building. If something wasn't working right,
I was the first one to notice."

But living right there in the workplace also had certain disadvantages. He and his family never really got a break from the job, and the lack of privacy was sometimes a challenge. "Visitors who didn't know the Centre would sometimes walk right into our living room, as if we were one of the exhibits."

The experience also taught Elliott the value of the volunteers. "Really tremendous people," he says. "If we needed a family break, one of the volunteers would come in and watch the place while we were gone. It's a bit of a sacrifice, giving up your Saturday night for Fort Whyte. But there was never a problem finding someone to pinch hit for us, and I was really impressed by that."

The volunteers had always been so effective that Fort Whyte had so far been able to operate almost exclusively on free labour. But that era was definitely over. With the expansion, the new interpretive building, and the multi-million dollar mandate, the Centre would need a full-time, paid management team, and Elliott's next move was to look for a site manager – someone who understood wildlife, could work well without supervision, and was handy with animals and birds and vehicles and machinery. Among the names of possible candidates was a guy named Ken Cudmore. He was a graduate of Lethbridge Community College's Wildlife Management & Resource Enforcement program and had been working as a jack of all trades in the wildlife business, banding ducks and doing other short term projects for the Manitoba Natural Resources Department. Elliott found his resume on file and called to see if he'd be interested in coming in for a chat.

Cudmore remembers getting the call. "I had put in my name at the old Fort Whyte Nature Centre about a year before so I guess they still had my application."

Site Manager, Ken Cudmore, on one of his annual spring "Goose Walks"

Cudmore came in to conduct a formal interview with the new CEO of Fort Whyte. "It was a conversation that was pretty typical of the way that Elliott and I relate to each other," says Cudmore. "It was about an hour long and I think we spent most of the time talking about dogs and duck hunting. When we were finished he said, 'Well, can you start Monday?'"

Elliott remembers it differently. "Ken has this blunt, straightforward manner that takes a little getting used to but is actually one of his best traits. The visitors love him because he's like an old-time backwoodsman. He just lays it on the line. We had our interview and I told him I would think about it, and I'd get back to him. Then a few days later he showed up in my living room and said: "So, have you made your decision?"

He got the job, and has been at Fort Whyte ever since. Apart from having what many Winnipeggers might consider to be the best job in the city – working with birds and fish and animals all day long, riding around on his trusty all-terrain vehicle, herding bison, planting trees, building nesting boxes, trapping skunks, doctoring injured geese, bottle-feeding newborn fawns – he has no boss, unless one counts Bill Elliott. "Ken just does his own thing," says Elliott. "He's like Daniel Boone. He knows more about this property than anyone, and he's never lost that rough charm that endears him to the visitors and the media."

Cudmore's help was much needed around the property. Elliott was swamped with work. "The building wasn't even finished yet," Elliott says. "We had no exhibits, unless you count Garry Hilderman's wall-mounted colour renderings of the Waterfowl Gardens that hadn't yet been built. The grounds were a mess. We hadn't done any landscaping or cleared any trails or pathways. We had less than a dozen volunteers and needed to recruit and train many more. So our start-up year was a bit chaotic, to say the least."

Amidst all the confusion, someone neglected to point out that it was necessary to clip the wings of the Centre's precious flock of captive geese, or their flight feathers would grow out after the moult and they would fly away. To complicate matters, the Centre inherited 25 wild Canadas that summer, all of which could fly. Ken Cudmore wasted many an afternoon chasing these new wild geese around the property, trying to catch them and clip their wing feathers so they wouldn't take off when cold weather arrived.

The geese eluded him, and in mid October the entire flock took off and disappeared. This was serious. Overnight, the old waterfowl refuge of the new Fort Whyte Centre of Environmental Education had no waterfowl. Through the winter, Cudmore and Elliott worried that the geese wouldn't return. Then one bright winter morning in March, Elliott looked out at the lake and the geese were standing on the ice, waiting for their morning handout.

Crown Vetch

Goose nesting raft

It was a relief to have the geese back. One of Cudmore's new responsibilities was to encourage them to stay by maintaining and enhancing the attractiveness of Fort Whyte as goose nesting habitat. He studied the birds as they waddled around the property. He needed to figure out what made them tick. The paired-off couples launched the new season by patrolling the property and inspecting likely-looking nesting sites. When they selected a place they shooed away their juvenile offspring and got to work building their nests. Having been kicked out of the house, the juveniles banded together with other one and two-year old "non-breeders" waiting to embark on the 'second migration' north to the Interlake in early June, where they would spend the summer molting and flightless on the comparative safety of larger waterbodies. Meanwhile, the adult geese then laid their eggs, but during the nights, foxes and other predators cleaned out the nests with surprising efficiency.

"I decided to help the geese by building five nesting rafts," says Cudmore. "Each one was equipped with a wooden crate packed with straw. I towed the rafts out to deep water, anchored them, and the geese immediately claimed them as nesting sites. That summer, the new nesting islands produced 19 goslings."

This was a triumph for both Cudmore and the geese, because predators continued hitting the onshore nests and not a single nest on shore was successful. The ducks weren't faring much better. "One pintail hen laid six eggs but a mink ate five of them. I patched the hole in the wire fence but the next night the mink returned, ate the remaining pintail egg and then killed a cinnamon teal and ate three of her five eggs."

Cudmore took the remaining eggs to the University of Manitoba and placed them in an incubator where eventually they both hatched. Meanwhile back at Fort Whyte a blue-winged teal laid six eggs and the mink ate all of them. Then a wood duck nest was destroyed by a raccoon. The mallards were having a hard time too. Only one mallard on then property managed to successfully raise a brood. "I inherited a wide and varied population of predators on the property," says Cudmore. "And until we installed a good variety of elevated nests it was a losing battle trying to outsmart them."

Some observers might argue that the conflict between predators and prey is a natural one, and the animals should be left alone to find their own equilibrium. But humanity has already interfered with nature to such an extent that nesting boxes are just an attempt to draw the pendulum back a bit. Predators take advantage of human development and incorporate it into their hunting strategy. Foxes and coyotes for example, hunt along farm fences and windbreaks, knowin that birds will build their nests there because most of the dense natural cover has been converted to cropland or pastur Foxes use the shorelines of Fort Whyte to similar advantage.

A female Canada goose must sit on her eggs for 28 days, an the eggs are so vulnerable in her absence that she will only leav the nest for a few minutes a day. (This is called "recess." Some geese have been observed to actually starve to death during this epic nest-sitting project.) When the young are hatched, th parents must be particularly vigilant, because there are countle animals in a wetland environment that view the goslings as plump, downy *hors d'oeuvres*. When you see a family of geese, you can observe this fierce, wary attitude in the parents. On th water, one parent will often lead the convoy of goslings while t other brings up the rear. And on land, the adults will lower th heads, spread their wings, and hiss at any human or four-foote interloper that dares to approach.

GOOSE LANGUAGE

Over the years, biologists have learned to interpret some of the bodily gestures, squawks, and honks of Canada geese. If you meet a goose on a walk through Fort Whyte, here's what he (or she) is trying to tell you:

- Pumping the head up and down:
  "You're annoying me."
- Lowering the head and hissing:
  "Now you're really annoying me!"
- Waggling the head from side to side:
  "Honey, you're looking good!"
- Repeatedly dipping the head underwater and splashing the back:
  "Let's get it on."

May 11/12 2004

The timing could not have been worse for our nesting Canada Geese. A freakish late spring snowstorm dumped 31 cm of heavy wet snow on Fort Whyte over a 48 hour period. Just days away from completing their 28 day incubation, the birds found themselves literally pinned to their eggs in a desperate attempt to maintain the 99.5 F temperature required by the developing embryos within. Despite the worst that Mother Nature could throw at them, the tenacious birds prevailed and another very successful Canada Goose hatch began just days after the storm's passing!

By the following year, Fort Whyte had a selection of floating and elevated platforms. "I was encouraged by the success of the nesting rafts," says Cudmore. "So I decided to build more. But geese are territorial, so I wanted to spread the nests out as much as possible."

He put some straw-filled boxes on poles mounted in shallow water, and his efforts at building and installing the meter-high nesting roosts were observed with interest by the geese. In one case a pair of geese stood on the shore watching Cudmore as he waded out into the water and installed a nest. "As soon as I was finished the female honked excitedly, and swam around the platform, examining it from every angle. She then jumped aboard, rearranged the straw, and settled in. A month later she successfully hatched six goslings."

The geese by now had become so accustomed to the security of the artificial nesting sites that they refused to nest on shore. And the proof was in the production. Formerly, predators had destroyed virtually every nest on shore. In 1985, thanks to the artificial nests, the resident goose flock produced 44 goslings.

GAZILLIONS OF GEESE

Coincident with Fort Whyte's development as a goose refuge, the City of Winnipeg began to require developers to include a system of "retention ponds" within the city limits. Primarily designed to retain stormwater, the ponds also beautified the city's new neighbourhoods. These ponds proved attractive to wildlife, and many geese from the country decided to move to the city. Golf courses, parks and boulevards provided ample lush grazing opportunities, and a lack of predators allowed them to raise their young in peace. Now, during the peak of the fall migration, more than 170,000 Canada geese congregate or "stage" within the City of Winnipeg.

Many of them use Fort Whyte as their evening drop-in centre, but thousands of others make themselves at home throughout the city. On any autumn day, the claxon yelp of flying geese is a constant backdrop to city life, and you only have to look above the office buildings of downtown Winnipeg to see those graceful chevrons drifting by. Geese in fact are now so abundant they are becoming a problem in some places – fouling golf courses with their droppings and congregating in worrisome numbers around airports. Wildlife officials are trying to suppress their numbers by lengthening hunting seasons and allowing limited hunts within Winnipeg's Perimeter Highway. But in general, most people regard Canada geese as majestic birds that provide beauty and natural grace to the city, and more importantly, signal the change of seasons.

When the founding members of the Wildlife Foundation of Manitoba decided to embark on a daring plan to raise a small flock of Canada geese at Fort Whyte, little did they realize how successful they would be.

Bill Elliott on the Compost Demonstration Trail

Meanwhile Bill Elliott was working on his general management duties, that entailed doffing many different hats. He had to interview new employees, answer the phone, meet with the Centre's supporters, tradesmen and visitors, and do the bookkeeping every day. "My work as a waterfowl biologist hadn't really prepared me for any of this accounting work," he says. "Nowadays we have a full-time accountant, but back in the start-up years I had to struggle with it myself. I was trying to do everything from payroll to taxes, so I was grateful to have board member and Treasurer, Art Tooley, as a mentor. He set up the books and spent a lot of time showing me how to manage them properly."

Elliott says Tooley offered "shrewd counsel" on the financial operations of the Centre. And to ensure that every penny was properly allocated, he co-signed the cheques. Tooley's voluntary guidance allowed Elliott to spend more time working on the mission for which he was hired – to come up with dramatic programming that would deliver the environmental message to Manitoba's young people. When kids learn about things from a textbook or a classroom lecture, it sometimes goes in one ear and out the other. But educational research indicates that there is much better retention when students physically participate in the lesson and see the results first hand.

His Royal Highness, Prince Edward is aided by Jonathan Elliott in the planting of a white spruce commemorating his visit in July, 1990

Art Tooley

The 'Energy Tornado' created by Richard Osen (above)

Roberta Bondar visits the "Earth Observation Theatre," 1996

Fort Whyte Centre's theatre earns applause in Vienna

By Manfred Jager
Staff Writer

THE PRESIDENT of Fort Whyte Centre has given a presentation about the entre's new satellite earth observation program to a United

Elliott, in an interview from the Austrian capital, said the Canadian Centre for Remote Sensing in Ottawa as well as remote sensing service and research laboratories of the Government of Manitoba and the University of Manitoba helped Fort Whyte

Fort Whyte's popularity with teachers and students is based on the idea that it's best to let kids plunge right in and discover the physical world for themselves. In one early exhibit, Elliott's brother-in-law was building a cedar strip canoe so Bill invited him to work on it at the Centre so the kids could watch it take shape. In another case, Elliott arranged to have installed a stationary bike to show the students how energy is converted. "We wanted to make the point that solar energy gets converted into plants and animals. Humans convert those plants and animals into food. The food gets converted into human muscle tissue. Riding the bicycle, the kids used the energy in their muscles to turn the front wheel and power up a series of lights – the harder one pedalled, the more lights turned on. It was a good exhibit because it combined hands-on fun with information. And with one simple machine, it taught a very complicated lesson about the interconnectedness of all energy forms."

Energy, of course, is at the heart of many of our current political and environmental woes. Air pollution, caused by automobile and industrial emissions, adds to global warming through the greenhouse effect. Wars and political upheaval can often be traced back to the oil supply and long-standing disputes over reserves. Water use, transportation, and even the size of our cities are all directly influenced by the availability of energy. So Elliott felt that it was appropriate that Fort Whyte should focus on energy, and over the last 25 years – ever since Manitoba school kids experimented with that first bicycle exhibit with its glowing lights – energy conservation has been a recurring theme at Fort Whyte. "We have had a long and very productive relationship with Manitoba Hydro," says Elliott. "These exhibits and interpretive programs can be quite expensive to mount and Hydro has been very generous with ongoing funding and technical expertise."

Elliott says that Fort Whyte's first duty is to the environment and education, so there is always the potential for conflict when corporate sponsors are involved. But he points out that Manitoba Hydro, under the leadership of its CEO, Bob Brennan, has been scrupulous about supporting energy conservation programming with no strings attached. "We could never jeopardize our credibility by shaping a message to suit our partners," says Elliott. "The costs would severely outweigh the benefits and it would be unethical on a variety of levels. But Manitoba Hydro has been hands off from the start. We've run exhibits that demonstrate the various costs and benefits of energy sources like solar, geothermal, and wind turbines, and although these alternative sources are sometimes in direct competition with hydroelectric power, Manitoba Hydro has been very supportive of the idea that people should know as much as possible about all energy alternatives and be as conservation-minded as possible."

One of the most dramatic exhibits was an "energy tornado" fabricated by dioramist, Richard Osen. With a chain link fence acting as the substructure of the vortex, he crafted a tall funnel cloud that was completely covered with both essential and frivolous energy-consuming items: portable radios, electric typewriters, sock driers, lava lamps, an automobile engine, electric blankets, and so on. "Some were almost obscene, like electric sock warmers, in their waste and lack of utility," says Elliott. "And others were devices we need and use every day, like toasters and water heaters. People would come in and stare at this great towering storm that was sucking up every kind of fuel and electric power imaginable, and it really got them thinking about the extent of energy consumption in their own lives."

In another alcove they created a hydro-electric dam diorama, complete with operating turbines, running water, and tiny sturgeon holding in the current downstream. "People just love these exhibits," says Elliott. "But it was all about education. We were the first depot in the province for newspaper recycling. We gathered the newspapers here and Kleysen's hauled them away. Then we started a glass and aluminum recycling depot but that didn't work out because some people abused it and we ended up with a wasp-infested garbage dump on our hands. So we dropped that… live and learn."

Fort Whyte also became the first place in the province to recycle Christmas trees. "There was a time in Winnipeg when if you were driving along in January and saw a used Christmas tree on the roof of a car it was almost certainly headed for Fort Whyte. We had mountains of Christmas trees."

Ken Cudmore would bind the trees together in groups of twenty or thirty, weigh them down with concrete blocks and sink them in the lake to create clumps of fish habitat or 'structure'. "Our lakes bottoms are fairly featureless and forage fish like to have cover where they can hide from predators. It worked really well for a few years, and the fish population seemed to increase. But then a scientist from the University of Manitoba called us and said we might be creating a problem with mercury. Scientists were just figuring out that submerged trees release mercury into the environment as they break down. So the educators got educated, and we ended that project."

Mountains of used Christmas trees are recycled as landscaping mulch, or bunched and sunk in Lake Devonian, providing shelter for its small forage fish

With Elliott's visioning and the influence of Maureen Krauss' management instincts, Fort Whyte established a pattern of pioneering various environmental initiatives, getting them up and running, then handing them off to other agencies and moving on. For example once the recycling business became mainstream Fort Whyte handed it over the city and went looking for other jobs that needed doing. "We weren't interested in getting into the long term recycling business," says Elliott. "Fort Whyte's role is to serve as the initial catalyst to get the public engaged in environmental stewardship."

Unlike a government agency, or an education facility funded primarily by the government, Fort Whyte's private status allows it to improvise and change programs with ease. It can quickly change its colours, adapting to new opportunities in its habitat. As Elliott says, "It's critical in the non-profit sector to be nibble. We keep trying different things. Most work, some don't, but that's all part of staying on the leading edge of education."

People sometimes ask Elliott about his "five year plan," and he has to admit that he doesn't have one. "We're in a constant state of evolution," he says. "We try to stay light on our feet and do our best to exploit whatever educational needs and opportunities come along. That's what we've learned from Nature."

Tonnes of newspapers were collected at Winnipeg's first recycling depot

Post Card from the Past

As the CEO of Fort Whyte, I've had two strokes of genius.

One of them was to apply for this job, and the other was to hire Maureen Krauss.

Around 1993 a vacancy arose in our fund raising office. So I placed an ad in the paper and looked at some applications and scheduled some interviews. One day I showed up early to conduct an interview and let myself into my office, but I forgot to unlock the front door. An early May snow storm was underway and at some point I looked outside and saw this poor woman trudging around in the snow, trying different doors, so I rushed out and let her in.

She walked in and shook off the snow and we got down to the interview. What can you say about Maureen? She's just so full of health and positive energy. She radiates so much enthusiasm, grace and charm that I just kind of sat there nodding, listening to her responses, thinking to myself: 'Fort Whyte has hit the jackpot'. So to make a long story short, I went through the motions and interviewed a number of people but the truth is, Maureen had the job early on in the interview.

She's now the second in command at Fort Whyte and I couldn't operate this place or make a decision without her. If I'm proposing an idea I can tell by her body language if she thinks I'm off track, and she's seldom wrong. If she embraces an idea then I know it's a good one. She gives this place class and style and she works like a dog. A couple of hours before a fund-raising dinner she'll be out there with a rag and a pail, cleaning chairs or whatever, sweat on her face. Then just before the guests arrive she'll duck into the washroom and come out looking like a different person, transformed from charwoman to 'charmwoman' in two minutes.

We're very lucky to have Maureen.

– Bill Elliott

CHAPTER 8

KEEPING THE WILD IN WINNIPEG

When Fort Whyte underwent its first major expansion in the early 1980s, it was essentially a waterfowl refuge. But since habitat conservation and broad-based ecosystem management are the best strategies for wildlife conservation, it seemed appropriate to adopt a more inclusive approach to other species. And nowadays Fort Whyte has evolved into Winnipeg's beacon for biodiversity.

The bison herd makes a big impression when visitors first enter the property. From spring to fall, many shorebirds, coots, and ducks cruise the lakeshore and cattail marsh. Hawks and gulls fly overhead. In the evening there are deer feeding in the fields. Crows, flycatchers, and sparrows are abundant. And if you look closely, you'll see evidence of creatures that are not quite so visible. Foxes, coyotes, skunks, raccoons, owls, and other opportunists work the night shift, and small rodents like voles and mice scurry through the undergrowth. All in all, it's a busy place. But it's much busier than most people realize. In fact, if visitors estimated the number of species inhabiting Fort Whyte they probably wouldn't even come close.

Dr. Bob Wrigley is one of Manitoba's most prominent wildlife biologists. He says that people tend to notice conspicuous animals like deer, foxes, and Canada geese, while overlooking smaller animals that are just as important to the overall ecosystem. "There are about 300 species of spiders," he says, "and 2,000 species of insects. Each one has been discovered and named by some intrepid scientist, but that's not necessarily true of the microscopic animals living in the water and the soil. There are literally thousands of invisible species in any complex ecosystem, and we're just learning about them. There are mysterious anaerobic bacteria living thirty meters underground and spiders riding balloons made of web material, flying high in the sky over the Interpretive Centre with the geese."

Prairie Skink

Wrigley says that Fort Whyte probably supports about 15,000 species. Some of those travel great distances in the spring and fall and depend on places like this to rest along their journey. The meadowlark that sings in the Bison Prairie on spring mornings flies all the way to South America in the autumn, where the antipodean summer is just beginning. The monarch butterfly that flits amongst the wildflowers behind the Alloway Reception Centre may be on its way to central Mexico, where logging threatens its mountainous wintering habitat. The yellow warbler that builds its nest in the aspen forest behind the Sod Hut has to run a gauntlet of nest-robbing predators like crows, raccoons and squirrels, all of which have exploded in numbers, thanks to the human civilization that furnishes them with abundant food and shelter. The lesser scaup ducks ("bluebills") that migrate through southern Manitoba are showing up in far fewer numbers on the choppy grey waters of Lake Devonian every autumn. Their dramatic decline continues to be the subject of much research.

Not long ago other species roamed Fort Whyte, and Wrigley likes to consider them too. "It gives you some historical perspective to think of the animals and plants that once lived here. I like to go back in time, imagining what this landscape must have looked like before human beings arrived. A little more than a century ago mule deer were the dominant species. And if you go back just two centuries, grizzly bears roamed this very patch of land."

Dr. Wrigley says that the field of environmental science is still new, and the information that scientists are gathering is alarming. "Scientists estimate that about a hundred species go extinct every day around the world," he says. "And the tragedy is, we know so little about these species that we don't even know what we're losing."

Years ago, when Bob Wrigley's wife died of a sudden illness, he brought his two young sons out to Fort Whyte the day after. It was the only place he could think of that could provide what they needed. "This has always been a place of healing," he says. "I think it's precious to many people. There are lots of Winnipeggers who aren't lucky enough to have a cottage or a place in the country. I don't know what they would do without Fort Whyte."

He says that he first developed the habit of seeking Nature when he was a boy, "Growing up in Buenos Aires my mother used to take me to the zoo, where I loved looking at the animals. It was fascinating to watch them and observe how their behaviour is so similar to ours. That's where I first got the idea that I would spend my life studying Nature."

Today, when he walks the trails of Fort Whyte, he observes the birds, trees and insects and finds that he is learning as much about himself as about them. "We tend to think of ourselves as being apart from Nature," he says. "Nature is over there on one side of the fence and we are over here on the other. From the time we're little kids we're told that mud is dirty, that spiders and snakes are dangerous, that Nature is an alien force that's out to get us. This creates an unfortunate gulf between human beings and the natural world. Many people think that if Nature wins, we lose, and vice versa. But we're a part of Nature. It's inside us. All these bacteria we're afraid of, the ones that you see people trying to eliminate with spray cleaners in TV commercials, are a beneficial part of the natural environment. They live inside our bodies by the millions, and we couldn't exist without them."

CHARLES ROBERT DARWIN.

POST CARD FROM THE PAST

A young peregrine falcon with a band on its leg spent quite a lot of time trying to catch a duck for breakfast. It would perch on the roof of the Centre and watch the mallards on the lake below. There were several hundred ducks on the lake and they were likewise watching the falcon. It's amazing to see how well the ducks can recognize a predator, even a motionless predator perched at a considerable distance. Peregrine falcons can only catch flying birds, so of course the mallards refused to fly, and for several mornings in a row I watched this staring contest between the ducks and the falcon.

One morning some newly arriving ducks flew into the lake. They didn't realize the falcon was on the roof. As soon as it saw them it launched into the air and sprinted towards them. Falcons are the fastest birds in the world and a peregrine can dive at up to 320 kilometres per hour. They need good altitude however to achieve high speed and this young falcon had only the roof of the Centre as a launching point. The ducks saw the approaching falcon and they simply folded their wings and plummeted like stones towards the safety of the water. Later on, the falcon dove at some gulls and they performed this same life-saving trick. It was a good demonstration of how difficult it can be for birds of prey to catch a meal. We tend to think of the falcon as being a super-effective predator, but this young bird was outfoxed time and time again. If it was so easy being a predator, we'd probably see a lot more falcons, and a lot fewer mallards.

– Ken Cudmore, 1990

Resident Wil

life

When he grew up, Wrigley considered becoming a surgeon but opted instead for wildlife biology, which doesn't pay as well but was closer to his heart. Over the years he worked for the Manitoba Museum of Man and Nature, as Director of the Oak Hammock Marsh Interpretive Centre, and is now the Curator of the Assiniboine Park Zoo, which he says is fitting, "given that I first became interested in animals when I was a child visiting a zoo."

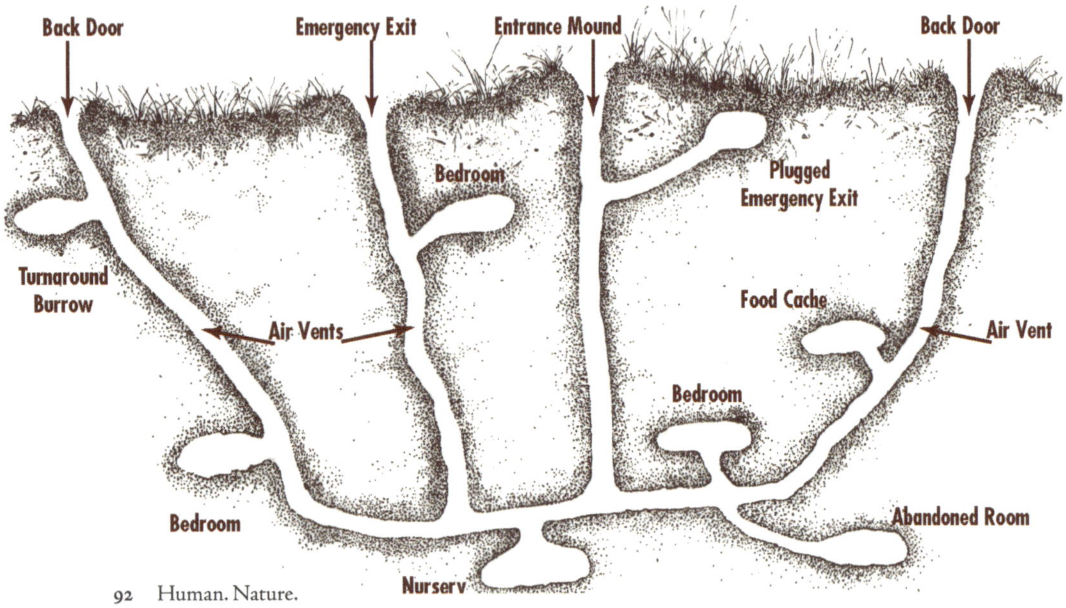

During the 1980s, he sat on the Board of Management at Fort Whyte, and that's when he and Bill Elliott discussed the possibility of installing an aquarium and a prairie dog exhibit. "I told Bill that of all the animals we had at the zoo, the monkeys and the black-tailed prairie dogs were the most popular," says Wrigley. "They are really delightful little animals. They're as fat as footballs in the autumn and they make these barking sounds, jumping in the air, and people just stand there watching them and laughing. There's always a sentinel on duty and if it gives an alarm call they all head for their burrows at full speed. After it rains, the burrows tend to collapse so they're always excavating earth, looking quite comical with dabs of mud on their noses. I like watching people watch the prairie dogs, because they get this look of delight on their faces. They become young again. It's amazing how Nature can do that to us."

It's fortunate that Fort Whyte provides a home for these animals, because populations of black-tailed prairie dogs have been severely depleted on the Great Plains. Ranchers dislike them because they compete with livestock for grass and their burrows can break a cow's leg. They have therefore been poisoned, shot and exterminated across most of their traditional range. But black-tailed prairie dogs are a keystone species. They are a critical food source for many other species like swift foxes, black-footed ferrets, burrowing owls, ferruginous hawks, golden eagles, prairie rattlesnakes, mountain plovers, and badgers. And as the prairie dogs have declined, all these other species have declined along with them.

We now know that prairie dogs not only supply these animals with food and shelter, but that the little rodents aerate the soil and help it retain moisture. A piece of Great Plains habitat with prairie dogs living on it has less flooding, richer plant growth, and more fertile soil. Prairie dogs, in other words, don't damage the environment, they help sustain it. And prairie dog towns are favourite grazing places for bison, pronghorn and even cattle. But as is so often the case, these lessons have been learned too late. Today, 98% of North America's prairie dogs have been exterminated, and the little dog town at Fort Whyte is the only colony in Manitoba outside the Assiniboine Zoo.

Establishing a prairie dog town at Fort Whyte wasn't straightforward. It's not enough to just fence off a piece of land and release a boxful of prairie dogs. The little rodents have to dig deep burrows to survive the northern winter, so it's necessary to bury fencing far below the surface or one day the dogs will pop from the ground outside the enclosure.

They also have to be housed with attention to their kinship. "Groups tend to be very competitive with each other over territory," says Wrigley. "They defend their borders fiercely, and they do that by sticking up for each other and behaving very aggressively towards outsiders. They touch noses and sniff each other and if they detect an intruder they'll chase it instantly and drive it off or kill it."

Fort Whyte used only inter-related prairie dogs from the Zoo, so the squabbles were minimal. And gradually the colony of Fort Whyte prairie dogs stabilized at its present level.

A Rare Sighting – "Our Moose"

A quick handed Lafarge Cement employee captures a photo of a bull moose on Lake Cargill's shore

POST CARD FROM THE PAST

I've had a great time watching a family of foxes this summer.

One sunny morning I spotted movement on the north shore of Lake Devonian. Close examination with the spotting scope revealed a litter of five kits, each one the size of a woolly football, romping at the water's edge. Not long after an adult fox swept onto the scene. The pups were obviously very happy to have Mom home. She soon had one pup pulling at her lip, another tugging at her tail, and the other three nipping and clambering over all parts in between. Her patience was amazing. She just lay there passively, offering up the occasional face wash with her tongue.

I was fortunate to observe the family on many occasions during the summer. As the pups grew older their boundaries for exploration spread further away from the den. Soon they were traveling several hundred meters down the shoreline, and by the end of the summer they were roaming all over the grounds. Many times I spotted them belly-crawling towards feeding geese. The geese would just retreat into the water, seeming unconcerned about the clumsy pups. It's a good thing that the Fort Whyte Centre has enough rodents, insects, frogs and other tasty items to eat, because the foxes would never have made a living from their goose-hunting skills.

Once they grew large enough to become independent of their mother the kits were still easy to identify. Instead of a rich coat of dark russet fur, they sported coats of dull yellow. The most notable difference was their trusting nature. Sometimes I would come upon them face to face and they would be more curious than fearful. One November morning I was snow-blowing the road and two of the kits were curled up sleeping in the ditch. Each time I passed with the tractor they did no more than lift their sleepy heads from the warmth of their bushy tails. They only thing they seemed concerned about was the racket that was disturbing their sleep.

Once the lakes froze, the waterfowl disappeared to a warmer clime, and with them went the carefree playful days of the foxes' youth. Winter is a long, severe season, and it's doubtful that Fort Whyte's 200 acres can support a family of foxes through the lean months. They will have to disperse, and wherever they go they will surely have to compete with other hungry foxes. While we lounge in the warmth and comfort of our homes, complaining about the -30 C. weather, the young foxes will be out there in the elements all winter long, trying to survive.

– Ken Cudmore, 1992

Tim Sopuck, Fort Whyte's first Director of Development, oversees the stocking of one of the lakes in support of the Young Anglers Program

CHAPTER 9 HUMAN. NATURE.

When the founders of the Fort Whyte Centre formed the Wildlife Foundation of Manitoba in 1966, they envisioned an organization supported by private citizens – farmers, city people, bird watchers, retirees, school children, scientists, and almost anyone else who appreciates nature and wildlife.

Ryan Porteous and Virginia Johnson guide school groups through diverse habitat

Over the years, it's true that Fort Whyte has been supported many times by various levels of government. Dedicated public servants like Gary Filmon, Sterling Lyon, Bill Clement, Brian Ransom, Bill Norrie, D. I. MacDonald, Sam Uskiw, and Boris Hryhorczuk ignored political expediency and threw their support behind Fort Whyte when the Centre needed them the most. But Fort Whyte was conceived as citizen-operated private entity, and to this day volunteerism and private sector support have served as its two cornerstones.

Some of these founding volunteers are deceased and some are growing old, but at one time or another they all carried Fort Whyte on their backs. Sam Fabro knew them all, and speaks of them with the laconic brevity of a Roman general remembering the fallen. There was John Baldner, "the champion trap shooter;" Richard Bowles, "our former L.G., a fine man;" A.R. Tucker, "a big shot at Eaton's;" D. I. MacDonald, "our general factotum, a senatorial type who was our source of sober second thought;" George Richardson, "there is nobody like him, and he has been on our side since Day One;" Major General Elliot Rodger, "a great human being, and deaf as a post."

Michelle Eldridge

"We had so many accomplished people," says Sam. "They frothed at the mouth for this place." As Fort Whyte grew, it needed more and more volunteers to deliver the educational programs and handle the countless tasks of servicing the public. They became the muscle and lifeblood of the Centre, and if it wasn't for them the organization would never have been able to stay in business.

At any given time, a small army of volunteers are onsite to help run the shop, answer the phones, greet visitors, take kids on nature hikes, and perform all the other myriad jobs that keep the operation on track. Visitors tend to notice the physical attractions of Fort Whyte – the wetlands, floating boardwalks, hiking trails, wildflowers, forests and wildlife – but only the volunteers can supply the energy and enthusiasm required to convey the environmental message to the tens of thousands of visitors who show up every year.

Who are these people? And what motivates them?

More than anyone else Michelle Eldridge should know. As the full-time Human Resources Manager, her job is to ride herd on a group of over 300 volunteers, all of whom have distinct roles and personalities and reasons for dedicating their time to Fort Whyte. Michelle understands these volunteers because she was a frequent visitor. "I first experienced Fort Whyte when I came here fishing with my son," she says. "Actually, I untangled line and he fished. After getting to know the place, I decided to volunteer, and that led to closer involvement, and then I was hired to work here. For the last twelve years I've been a full time member of the Fort Whyte staff."

Michelle grew up close to nature herself, in St. Vital on the southern edge of the city. Her earliest memories are of roaming the nearby forest when she was eight years old. "I would sit in the meadow and eat my cucumber sandwich, perfectly happy to be just exploring the woods alone. I guess kids don't or can't do that today."

She said she has always been interested in exploring new areas, and that seems to be the main motivation driving the hundreds of people who volunteer at Fort Whyte. "They're all very different people of course, and many of them come with a pretty impressive knowledge base. But I'd say the thing they have in common is a desire to learn."

She meets with each prospective volunteer and gets a sense of how she can match them to the needs of the Centre. "Many of them already have busy schedules. We respect that, and each person pretty much determines the intensity of their own involvement."

Her job may resemble that of an infantry major, but she says that she never has to lay down the law. "In my experience it's never been a concern that volunteers show up late or cancel at the last minute. Quite the opposite – they take their jobs very seriously. When they're all together in a group you see how enthusiastic they are. It's such an extraordinary place, with the birds and the wildlife and the trees that I guess it inspires people to do their best. When people realize the whole enterprise depends on them something magical happens."

Celebrating the end of another successful school year, Bill and Maureen flip burgers at the annual June Volunteer Windup Barbeque.

Every spring she organizes about 110 junior counsellors to help run the popular Day Camp program. "Somehow everyone gets woven into a seamless fabric," she says. "We have a special newsletter that goes out to all the volunteers so they know what everyone else is doing. People remain flexible and they work together and it becomes a real team effort."

She says that the young volunteers often discover their own career paths by helping younger people learn basic skills about the outdoors. "I love watching it. You can see these young counsellors discovering who they are. They try some teaching and they discover a set of skills they never realized they had. Before you know it they're asking for a letter of reference to the Faculty of Education."

At the other end of the scale are the grandparents and retirees, older people who have a bit of time on their hands and have realized that sometimes working for something you believe in can be more rewarding than working for a paycheque. "Retirement can offer the opportunity to reinvent oneself," she says. "For example I know one gentleman who was an aviation mechanic, and spent his whole life working with things that don't talk back, so his socializing skills were perhaps a bit rusty. But he started working with kids in our school program, doing wilderness survival classes and so on, and found out he was really good at it. He found this new part of himself that he had never developed and he just blossomed. Another lady spent her whole life in an office then came to us after she retired. Now she's the most amazing teacher. It's quite inspiring to see this happen to people. Quite a few of our volunteers will tell you they found whole new lives at the age of sixty."

It's impossible to give proper recognition to these volunteers – there are simply too many of them – but anyone who browses through the backlist will discover extraordinary people. Pick a name at random and you'll find someone like Don Willcox, who has donated 20 years of his life to Fort Whyte; Louise Staples, who drives in from the country three times a week to work full days at the Centre; young Jacob Kettner, who started volunteering at the age of 14, and when not entertaining younger students with feats of magic worked as a camp counsellor and enjoyed it so much he is now considering a career in Environmental Science. There are couples too, like Judy Owen and her late husband, Steve, who hatched the idea of the ever-popular Sunset Barbecue, and more importantly, handled the myriad organizational tasks that transform a good idea into a smoothly functioning event

Another couple, Bob and Judy Stewart, regularly put on pioneer clothing and welcome visitors to the rustic sod hut in the woods next to the Alloway Reception Centre. Bob, with his broad shoulders and grey beard, looks every inch the sturdy sodbuster, and Judy, baking biscuits inside the hut, is always full of good humour and information about the way pioneers lived in the 1800s. The walls and roof are, of course, made of prairie sod, and though the hut is pleasantly cool on hot days and cozy warm on cold winter mornings, Judy reminds visitors that life was challenging for a mom in those days. "You had to go all winter without a bath," she says. "And it was important to make the door swing inwards, otherwise you'd be trapped inside the first time there was a big snowstorm."

Don Willcox

Out in the yard, where a large Canada goose drowses in the sun, looking as unconcerned as the family dog, Bob stands next to the authentic steel plough that he and Judy donated. Hooking his thumbs in his suspenders, Bob never tires of spinning yarns about the old days. "The sod roof has its drawbacks," he admits. "If you have a big rainstorm, it keeps on raining inside the house for the next three days."

Bob says that his history with these woods goes well back into the 1950s. Long before the Fort Whyte Centre existed, he and his fellow Boy Scouts would undertake 10-mile hikes out here in the dead of winter, and tramp across the open fields south from Wilkes Avenue on snowshoes. "We would often see deer, cottontail rabbits, jackrabbits, and foxes," he says. "And we'd get very cold."

With their requisite one match, they'd start a campfire in the woods, upon which they would warm a pot of soup. "Hiking home was the hard part," he recalls. "It was usually a north wind at that time of year and we never seemed to remember that we'd have to march into it on the way home. Our clothes were damp and our faces were frozen but we'd keep our spirits up by singing 'I Found My Thrill on Blueberry Hill.'"

Bob says that on one snowshoeing trip they almost blinded themselves with the glare off the snow, so they did some research in the encyclopaedia and constructed Eskimo-style snow goggles with cardboard and string. In the summer they camped out here in pup tents, got rained on and devoured by mosquitoes. One morning they were startled when a huge animal stuck its head into the tent. They were relieved to see that it was only a cow.

Bob grew up to become a teacher at St. John's Ravenscourt, and during the 1970s, his class included the children of Alan and Helen Scarth. The Scarths invited Bob and Judy to get involved in Fort Whyte and they happily agreed. Alan Scarth needed someone to plan and cut some interpretive nature trails, and who better than the grown-up Boy Scout who had spent so many Saturday afternoons prowling these same wooded paths as a trespassing youngster?

With the help of two student volunteers, Bob scoured the woods for practical trail routes, cleared brush, built bridges, and set posts for signage. That fall, a professional interpreter was hired to introduce the public to the lakes, marsh, swamp, aspen forest and meadows along the trail system. To this day, hiking remains one of the most popular activities at Fort Whyte, and none of it would be possible without the spirit of volunteerism exemplified by the Stewarts and those young students. "It's been entirely our privilege to be connected to this place," says Bob. "What a joy it has been, both in my life and beyond, to know that Fort Whyte will always be here for my children and grandchildren."

UNTEERS

Volunteer Sod House Pioneers: Judy Stewart, spinning wool. Bob Stewart, spinning tales.

The volunteers are legion, and each one has a great story to tell. Terry Brownlee, another long-serving volunteer, grew up on a farm in Souris, where he dreamed of becoming a disc jockey. You couldn't get a job with a radio station without experience, so he enrolled in a radio operator's course. But it turned out that the course was about "high-frequency communication," not broadcasting, so he ended up becoming an Air Traffic Controller. It was a good choice. Today, as he walks the trails of Fort Whyte, he's always looking at the sky, predicting what the weather will be tomorrow. He can name the types of clouds, show you ways of locating the centre of an incoming system, (orient yourself into the wind and extend your right arm), and with a glance overhead, identify the make of a plane cruising thousands of feet overhead.

Heather Anderson working at the Visitor Reception desk.

Terry worked all across Canada, and ended up managing ATC operations at Winnipeg airport. But deep inside he was still a farm boy who missed walking the woods and fields of home. "I used to come out to Fort Whyte with my ex-wife and our three kids when there was nothing here," he says. "The dragline was still digging out the pit. We came out here to get away from the city for a few hours and walk in the woods."

When the Interpretive Centre opened in 1983 he volunteered to take school kids on nature walks. "I've never believed the common misconception that modern kids are spoiled and disinterested in nature," he says. "They're interested in anything that's interesting. It's our job to grab their attention. They're like sponges. They'll absorb everything around them. If they're not listening, it means that we're not presenting the session in a lively and engaging way. I have always tried to give it my utmost, as an interpreter, and I've never had a problem with bored kids."

After his divorce Terry kept up his connection to Fort Whyte. And as the seasons changed, from falling leaves to falling snow and back to spring again, he enjoyed the family atmosphere of the place. His kids were growing up, finding interests of their own, and Fort Whyte was like a home away from home.

Sometimes when arriving or leaving he chatted with Heather Anderson, the Centre's Receptionist and Booking Coordinator. Heather, like him, was divorced with grown up kids. She started working at Fort Whyte in 1985, when the building was still so unfinished that on cold days she had to work at her desk in a ski jacket. But Heather did her best, and somehow managed to get everything done. As Terry puts it, she was the Centre's "all-around Girl Friday."

Terry was extremely impressed with Heather's professionalism. "She had the answer to every question," he says. "No matter how chaotic things got with kids pouring in the door and phones ringing I never once saw her rattled."

He says he wanted to ask her out on a date but he didn't have the nerve. "I don't know how many times I dialled her number and then hung up before she could answer."

Then one day Heather decided she was going to take a well-paying job at the Manitoba Museum, one with actual pension benefits, and the news that she was leaving gave Terry the extra nudge he needed to make the call. "So I called her."

Heather picked up the phone. "It was Terry Brownlee," she recalls. "He told me he was volunteering for the Rainbow Society at the Red River Exhibition. He wanted to know if I wanted to come out to the Ex and go for a walk."

Heather paused. She didn't know what to say. "I didn't want to get involved," she explains. "Not that I hated men or anything, I was just happy being on my own. And I also hated the Red River Exhibition. It was so crowded and noisy, the last place in the city I wanted to go."

Terry weathered a long uncomfortable pause as Heather tried to figure a diplomatic way to explain this to him, then she thought, "what the heck, what's the harm in going for a little walk? So I agreed to meet him, and now we've recently celebrated our ninth wedding anniversary."

Just recently, Terry bought her a little surprise. If you walk through the swamp boardwalk just to the southeast of the Interpretive Centre, where on a typical summer day the kids are on their hands and knees dip-netting aquatic insects, you may notice that many of the planks in the boardwalk have little plaques on them. For a small donation, anyone can "buy" a dedicated plank in the boardwalk. The tax-deductible donation ($50) entitles the owner to one numbered plank with a choice of wording, a "certificate of ownership" and a complimentary family day pass to allow the owner to come and view their own personal part of this long meandering frog-chorus trail. Each plank in the boardwalk refers to a person/individual, and invariably that is someone who loves Fort Whyte. One plank in particular refers to the romance between Heather and Terry. It reads, "Heather – the love of my life, Terry."

This Giving Grove of mixed aspen and burr oak is dedicated to all past and present Fort Whyte Volunteers, celebrating forever their inspiring spirit and passion that keeps Fort Whyte vibrant for generation after generation of visitors.

You'll see benches along the trail too. The idea of getting private citizens to sponsor concert hall seats or park benches is not a new one, so most people might walk past the benches without giving them much thought. But each one of these benches contains the essence of somebody's life and their love of nature.

One bench, for example, is inscribed: *In Memory of David Plews – Nature Was His Passion*. David Plews was a chemist who worked for Manitoba Conservation. His job was to keep an eye on pesticides in the environment and respond to emergencies like derailments involving ruptured tank cars. David got a lot of satisfaction working to ensure that wild animals, birds and waterways were protected. His hobby was bird watching, and he always took binoculars with him when he was on the road in rural Manitoba. When he was at home in Winnipeg his favourite spot for birding was Fort Whyte. "He was always out there on the weekends," his wife Lia Wilson says. "And weeknights too. I went out there with him sometimes but when it came to birds he was such a keener that to tell the truth I had a hard time keeping up with him."

Lia didn't have much interest in birds until she got to know David, but his enthusiasm was contagious. And he not only won over Lia but also his children and eventually his grandchildren and even his relatives in the United States. Says Lia, "He got us all hooked on birding."

In 2004, David died suddenly of a heart defect. "It was very quick," she recalls. "He used to say that that was exactly how he wanted to go." Lia wanted to commemorate him in some way. "A bench out at Fort Whyte seemed like the perfect gesture. I didn't want to just send money to some charity. I wanted to do something that would be meaningful to David. He loved Fort Whyte and helped out there in so many ways. I liked the idea of installing a bench where people could come along and rest and watch the birds he loved so much."

Lia eventually purchased three benches, each one in a quiet vantage point along the trail. Each has a different inscription, and each affords a view of a different type of bird habitat. "For me the benches are a place to reflect on David's life and on nature," she says. "I'd much rather go to Fort Whyte than to a cemetery. A bench is a living memory and it continues to support birds and conservation and the things that were important to him."

David's late mother, in her 80s, appreciated that the benches are accessible to people with limited mobility. And David's grandkids come in from the small town of Ninette, Manitoba to visit Fort Whyte. "They sit on the benches, watch the birds, and talk about what grandpa taught them about the different species. Even our relatives from the States got introduced to Fort Whyte thanks to David. It's a very positive and pleasant way of remembering him."

Another bench commemorates Bud Malcom, the late owner of Malcom Construction. Bud was a well respected figure in the Winnipeg business community and one of those rare people who attracted nothing but praise after a long career in a notoriously contentious industry. He built two of the first buildings at Fort Whyte and invariably brought them in under budget and ahead of schedule. "When Bud was alive we had a place out at Oakbank where we could enjoy the country life," says his widow Joan. "But Bud had a stroke in 1981 so we moved to Charleswood to be closer to the city. That's where we became interested in bird watching. We had tremendous enjoyment watching the deer, the foxes and the birds. We often went out to Fort Whyte to go for walks. We'd take a picnic lunch and sit down by the water and watch the birds. It was really enjoyable."

A 13-acre orchard of several hundred apple, plum, pear, and cherry trees serves as another crop of Fort Whyte Farms and a unique opportunity to celebrate special life events or remember those who made life special, by dedicating a tree and sharing in its annual harvest.

To commemorate Bud, Joan donated money for a plaque and gave money to the school program, a cause which was dear to Bud. "He loved the letters the kids wrote about their visits to Fort Whyte. They were quite funny and quite touching, some of them. We often take wildlife for granted but some of the kids didn't know what a deer was. He felt it was so important for urban kids to have a place to go and learn about nature."

Bud's sister Allison Molgat, widow of well-known Manitoba politician and former Senator Gil Molgat, says that Bud, a father of four, was a devoted husband, brother and family man with friends galore. "He had a cottage at Clearwater Bay and I remember he would make his coffee in the morning and go down and sit on a bench by the water and watch the loons and the morning sun on the lake. He was never discouraged about life. For Bud, it was always a sunny day."

Like his widow Joan, his sister Allison wanted to commemorate Bud and his connection to Fort Whyte. "It's really Joan who made the major contribution, but I wanted to make a small gesture in my own way, so I purchased a bench in a secluded place along the pond. It's a tranquil corner, definitely his kind of place," she says. The inscription on the bench reads, *"Bud Malcom – He so loved this wild and peaceful place and all whose home it is." – Love, Allie and Gil.*

If you take a walk along the trail running northeast of the centre, past the treehouse and the bison viewing mound, you come to a truly secluded place called Lake Cargill. Surrounded by forest, Lake Cargill looks like it was scooped out of the Interlake wilderness country and dropped into the middle of Fort Whyte. You wouldn't be surprised to see an elk walk past. On the east shore of Lake Cargill is the Siobhan Richardson Field Station. It's an impressive looking gathering of buildings, like a kids' summer camp, with bunkhouses, a swimming platform, and canoes and kayaks pulled up on the shore. Most people walking past look at it and assume that it's just another part of the Centre. But like every bench and plaque on the property the "Siobhan" has a story to tell.

When Joan Rankin married James A. Richardson (the fifth Richardson to carry the JAR name) she brought her own skills and ambitions to that famous Manitoba family. "I have always volunteered and I've always been outdoorsy," she says. "I used to run the Therapeutic Riding Association and I'm quite an experienced duck hunter. I even have my own shotgun. It's a 16 gauge, quite a rare and pretty little gun but it's very hard to get shells for it."

She met Bill Elliott out at the Delta Waterfowl Research Station north of Portage la Prairie in the early 1980s and through him, started volunteering at Fort Whyte in 1984. "I was on the very first Sunset Barbecue Committee, about twenty years ago. We were on a real learning curve. There was the famous hurricane barbecue, for example, when we were afraid the windows were going to break and the wind blew a porta-potty way up in the air and onto a trustee's car. Most of the food didn't get eaten because it was blown all over the place under a collapsed tent. I think the racoons were the biggest beneficiaries of that event. But we had better luck in other years and it has turned into a really great fund raiser. There's great food, a good crowd and it usually brings in about fifty thousand dollars."

She also spent nine years volunteering as an interpreter with the school groups, simultaneously raising a big family. "I was often pregnant and sometimes I couldn't fit in the vole tunnel."

Along the way, she fell in love with Fort Whyte. "It's like no place else. You have such high quality people there. I'm just this little mom with an urban planning degree, but the guy who's leading the field trip might be a doctor or a scientist, and I find it very rewarding to be working with that calibre of volunteer."

She also became a board member, and that experience gave her even more respect for Fort Whyte. "This place does more with its money than any place I've seen, and I've served on a lot of boards. It's a lot of fun, working here. I'm sure there's not another board room in the country where you have to concentrate so hard on what people are saying because two geese are having a big squabble outside the ground-level window. The most fun we have is with Bill Elliott, the poor guy. No one misses the board meetings because they are so entertaining. There is always lots of work to do, but we have a great time doing it."

In 1985, Joan gave birth to a baby girl named Siobhan, who only lived for two weeks. As a way of coming to terms with that loss, she and Jim started a charity in the baby's name – The Siobhan Richardson Foundation. They gave the hospital an incubator for premature babies. And they decided to devote another package of funding to build a field station facility at Fort Whyte for older kids.

"Joan has always been a great source of ideas," says Bill Elliott. "And we've been through a lot of highs and lows together. When she started the Siobhan Richardson Foundation we got onto the idea of creating a massive floating building for an educational facility. It would have portholes down below so you could see underwater and a lot of other features that were very cool but it turned out to be just unworkable because of the cost. Also, what do you do with a building like that in the wintertime? It would be too big to take out of the water and you'd be worried about ice damage. In the end, we decided to go with a land-based operation right on the shore of the lake."

Annual Sunset Barbecue Fundraiser

Fort Whyte Board of Trustees & Council Members (L to R)

Rob Simpson, Joan Richardson, Harry Duckworth, Bill Parrish Jr., Michael Scatliff, Timothy Kurbis, George Richardson, Thomas Henley, Douglas Harvey, Fred Monnin, Ruth Krindle, Sam Fabro, Alan Scarth, David Grant, Malcolm Crooks, Dave Burrows, Michael Susser, Bill Elliott, Bill Mitchell

The Siobhan Richardson Field Station sits on the east shore of Lake Cargill. There is a road, (to provide access for disabled participants and emergency vehicles,) but in general the students must come and go on foot, by bicycle, snowshoes, or some other means that doesn't use fossil fuels. School groups of various ages practice outdoor skills and in some cases camp overnight in bunkhouses that were built by student volunteers from the University of Winnipeg Faculty of Architecture.

The educational program is directly geared to the science curriculum so that kids can work with ideas they have studied in school. On the "climate change" theme they are exposed to energy sources such as wind, wood, geothermal, solar and hydroelectric, studying the costs and benefits of each. It has become a sort of mini-university of environmental learning, and it's a fitting legacy to a little girl who will never have a chance to study there.

"We wanted to commemorate a life that would never be lived," says Joan Richardson. "Siobhan couldn't grow up and work here as a volunteer herself, so the Field Station is her contribution to Fort Whyte."

As visitors walk around the grounds or ski along the trails in the wintertime, they always notice the deer and the birds. Those who look closely also find stories hidden in the planks of the boardwalk, the benches along the hiking trails, and in installations like the Siobhan Richardson Field Station – reminders of the community spirit that founded Fort Whyte.

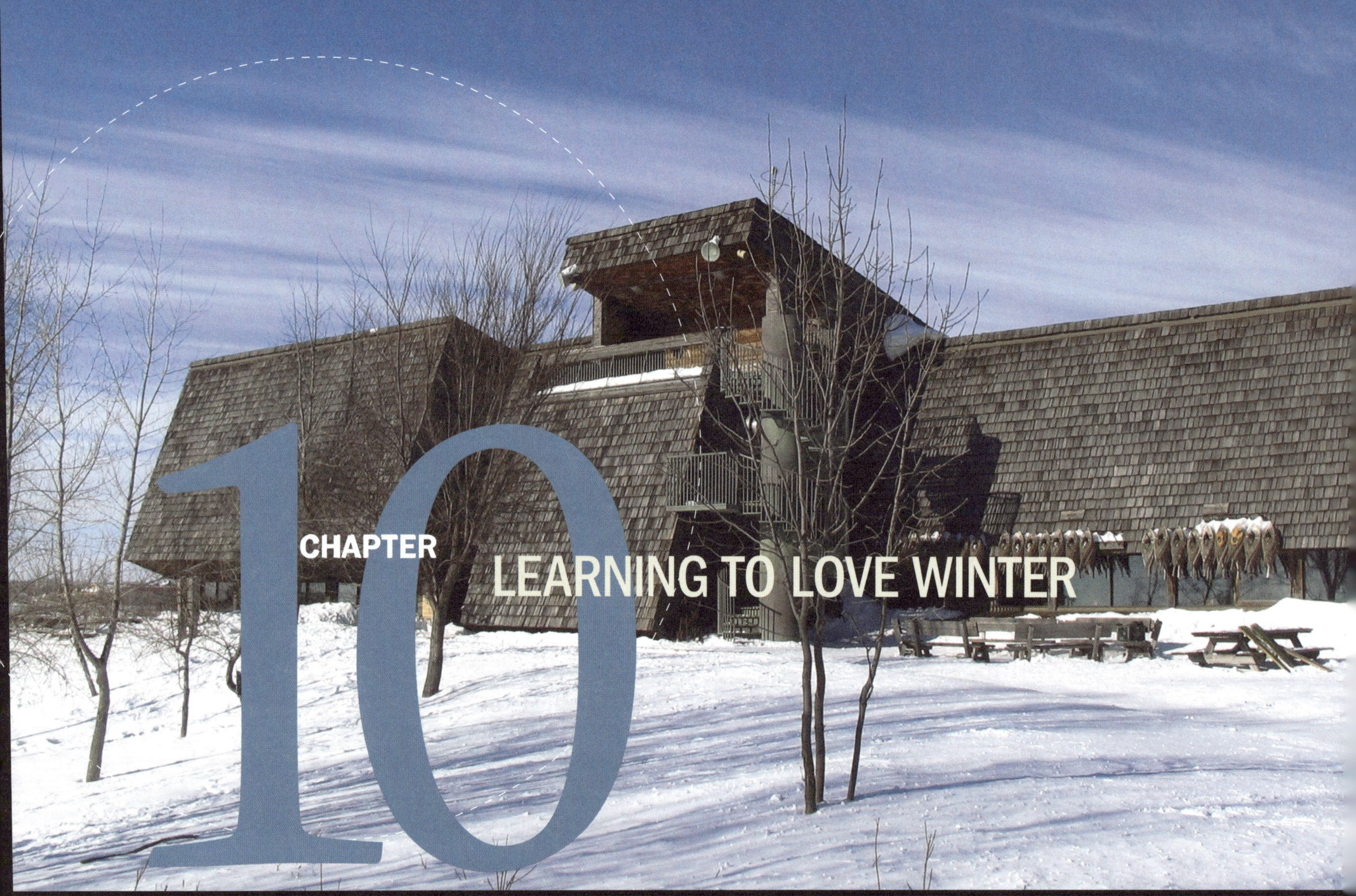

CHAPTER 10 LEARNING TO LOVE WINTER

In the early days of Fort Whyte, senior staff realized they are actually involved in two different businesses – the environmental education business and the outdoor recreation business.

Environmental education takes place on week days, when yellow school buses roll through the main gates and disgorge hundreds of school kids. With the help of teachers and volunteer interpreters, those schoolchildren spend hours looking at the wildlife, viewing the exhibits, and learning about Nature. On the weekends, Fort Whyte changes hats and becomes a recreation centre, a place where families, kids, older adults, tourists, lovers, and many others come to walk the trails and indulge in some much needed contact with peace, beauty, and unspoiled Nature.

The education business was, and continues to be the heart and soul of the Centre. But that business is a money loser. There is very little leeway in the amount that Fort Whyte can charge the schools and students. School Divisions have limited budgets for field trips and bussing, and the fees paid by the schools or parents are not adequate to cover the costs of educating each student, so the Centre is left with a significant shortfall, which has to be recovered from the recreation side of the business and donations. Government grants cannot be relied on for ongoing operations. So in order to fulfill its mission – to educate youngsters on the importance of environmental stewardship – Elliott realized at a very early stage of his presidency that the Centre needed to get into the recreational business and somehow attract more paying customers, especially on the weekends.

Manitoba, as we all know, is an easy sell in the summertime. Thousands of tourists from the United States, Asia and Europe come to Manitoba every summer to explore the city and enjoy the outdoorsy diversions of rural and wilderness Manitoba. But those numbers drop off drastically once the autumn winds start to blow, and even native Manitobans hibernate indoors during the long months of winter. Bill Elliott, Maureen Krauss, Ken Cudmore and the other members of the Fort Whyte team tackled this dilemma by creating a number of attractions, programs, and exhibits designed to lure people "off their couches" and out to the Centre on winter weekends.

The waterfowl, resplendent in their breeding plumage, were once housed behind glass, but thanks to the Richardson Foundation a new viewing deck allows people to go right into the waterfowl room and watch and hear the birds preen and cavort. The Richardsons helped build the waterfowl viewing deck and financed the construction of a toboggan slide, "The Richardson Rrrrun," which has turned out to be very popular. On the coldest winter day, people line up to ride the slide, and not all the keeners are kids. "I go down it every year," says Maureen. "And it's quite exhilarating. But I always wait until Ken tests it out."

Since the water level of Lake Devonian varies from year to year, Cudmore has to build a transition ramp that carries riders from the toboggan slide out onto the frozen lake. He tests the angle of the transition ramp by being the first one to try it. (He's considered to be not so much expendable as unbreakable. But he's a good enough carpenter that so far it hasn't been an issue.) After getting good and chilled outside, the toboggan riders can come inside and warm up with a bowl of bison chili or a cup of hot chocolate in the Buffalo Stone Café.

"Before we had the Café, people were always asking where they could get a cup of coffee or a bite to eat," says Elliott. "So when we planned the Alloway Reception Centre a restaurant was one of our priorities. It's proven to be very popular. A lot of people come out here just to have a bowl of soup or a Bison Burger and watch the kids sliding onto the lake."

SNOW

Snow is both a burden and a blessing for the wild creatures that overwinter at Fort Whyte.

White-tailed deer for example depend on plants and grasses that are covered up by the first snow. They scratch away the snow cover with their hooves, but as the snow deepens they switch to browse – plants and buds that are tall enough to rise out of the snow. Browse is not sufficiently nutritious to support deer in the long term. So from midwinter on, they start losing body weight and it's a race against the calendar. During a long cold winter, many deer in remote areas of Manitoba will starve before the greens of spring arrive.

Snow reflects up to 90% of the sunlight striking it, thereby cooling the ambient air and prolonging our winters. But it also has a low thermal conductivity so it is a first-class insulator. Rabbits dig into the snow to insulate themselves from the cold, and mice spend most of the winter beneath it, building little tunnel-cities, where the temperature is never more than a few degrees below freezing.

Without the "subnivean climate" that exists under the snow, the lakes and small pothole sloughs that dot the Manitoba prairie might freeze to the bottom, killing the muskrats, turtles, frogs and other creatures overwintering under the ice. And snow protects the myriad plants and insects that huddle under its protective blanket. Animals like prairie dogs, ground hogs, and bears put on weight during the summer months, and survive the winter by holing up underground, beneath a frost line that is directly influenced by the snow above. Snow even helps these animals after it melts. One of the first foods that black bears fill up on when they wake up is *Hypogastrura nivicola*, the snow flea. The forest looks like a barren place in early spring. But snow fleas manage to survive the cold thanks to an antifreeze-like substance in their bodies, which allows them to roam the surface of the snow on sunny winter days and to congregate in vast numbers in matted leaves under the snow. In spring, bears shuffle along with their snouts in the leaves, lapping up these tiny insects by the bushel.

For Manitobans, snow is likewise a blessing and a curse. We toboggan on it, pay millions in taxes every time it falls, strain our backs as we shovel it, and get dewy-eyed and sentimental as we watch it fluttering down through the woods on a winter evening. On the prairies, snow is such a ubiquitous part of our lives that we often forget to see it for the remarkable substance it is. But snowflakes have always fascinated scientists. In 1611, Johannes Kepler published a not-so-famous treatise entitled *The Six-Cornered Snowflake*. Today, snowflake morphology is studied by scientists from all over the world, who sometimes cite Manitoba as one of the best places in the world to find beautiful flakes.

One of the most persistent beliefs about snowflakes is that "no two are alike." This bit of folklore was popularized by a self-educated expert from Vermont named Wilson "Snowflake" Bentley, who in 1885 attached a microscope to a bellows camera and became the first person to photograph a snowflake. He eventually obtained pictures of over 5,000 flakes, and his observation that no design is ever repeated passed into the accepted record. Was he right? No one really knows. Standing outside at Fort Whyte on a January evening, watching thousands of tonnes of snow descending on Lake Devonian, it might be a little hard to believe that no two of those flakes are alike. The sheer number of snowflakes in even a modest Manitoba snowdrift has so far deterred any researchers from getting out their microscopes and disproving Bentley's thesis. So there's a good project for any of you folks with time on your hands.

Winter is A
at Fort Wh

live
yte

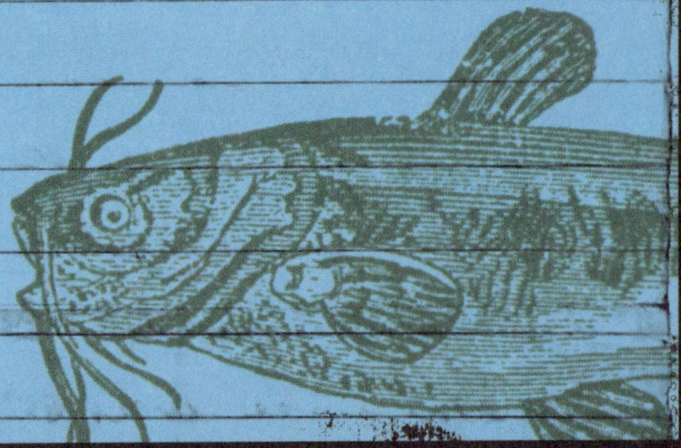

RED RIVER AQUARIUM
This 9,000 litre aquarium represents
Typically shallow and warm, these
fertile and host a complex food
a large and diverse fish population.
Manitoba's prairies are classified
Examples are: Assiniboine River,
Lake Manitoba, Lake Winnipeg.

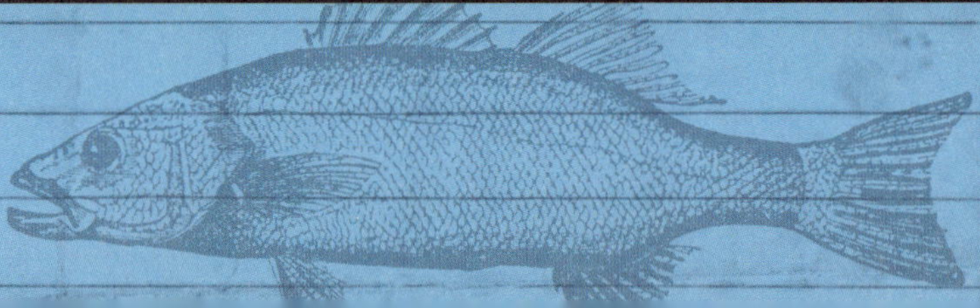

Another popular winter attraction is the Aquarium of the Prairies. Built in 1989, the aquarium was an enormous undertaking, necessitating an entire reconstruction of the building's south wall and considerable foundation work. One component of the aquarium holds 16,000 gallons and represents a lake in Manitoba's Precambrian Shield. It is 12 feet deep, with jackpine snags, simulated granite walls and a lake bottom of rock rubble. Pike, walleye, bass and other typical shield country fish cruise spookily through the gloom, waiting perhaps for Cudmore to arrive with a bucket of live fathead minnows that he's trapped from one of Fort Whyte's lakes. The second smaller tank is 2,000 gallons and represents a shallow beaver pond. Children are fascinated by both tanks, and it's always quiet in the viewing room, with many small noses pressed up against the glass.

(Above L) Aquarium of the Prairies under construction

(Above R) Artist and sculptor, Richard Osen, uses an eroded shoreline of Lake Devonian to cast the fibreglass underwater background in the Aquarium of the Prairies exhibit.

Ken Cudmore is in charge of keeping the tanks clean and populated with fish, turtles and other aquatic creatures. "Like everything, getting the water in and out of the aquariums was a real learning curve for us," he says. "We're always reusing and recycling here, so we use water from Lake Devonian, and we went through a lot of trial and error, trying to figure out how to circulate that much water without constantly clogging the filters. It's never been a problem getting fish for the tank. With a collection permit from Manitoba Conservation I'll go out to the lake country on the weekend and catch them with rod and line. I don't mind that job!"

The Kiwanis "Touch Museum" has also turned out to be a sleeper hit. As Bill Elliott explains, "We got the idea from those 'Please Don't Touch' signs you often see in museums. I wanted to install a wildlife exhibit where the kids were encouraged to 'Please Touch.'"

With its menagerie of mounted bears, hawks, beavers and other wildlife, the Kiwanis Touch Museum is perhaps the most popular exhibit after the aquarium and the live waterfowl display deck. "Kids just love to get in there and pat those animals," says Elliott. "If the exhibits get a little beaten up after a while, that's okay, we just replace them, opportunistically of course, with a mounted road kill specimen or a periodic donation from someone's cottage or recreation room."

The Prairie Soils Diorama is more educational than entertaining. It's true that there's nothing glamorous about dirt. But soil is a fascinating topic, once you dig into it, and Elliott decided that since agriculture is a major sector of Manitoba's economy and an environmental force on the landscape, the Centre needed to focus on the often-ignored subject of soil conservation. "The topsoil we depend upon for food production took 10,000 years to create," he says. "But we've degraded much of it in only a hundred years. Respected scientific bodies like the Worldwatch Institute have determined that we're losing tens of billions of tonnes of topsoil a year through wind and water erosion. In a world where people are starving to death, we can't afford to lose any productive land."

The Year is 2050
You can take action on climate change.

Acclaimed dioramist, Betsy Thorsteinson, at work on the Prairie Soils minature diorama, 1991

There are many causes of soil erosion. Huge sections of Africa have been overgrazed and turned into barren desert by domestic goats. In South America, small farmers burn down the rainforest desperately seeking a few years of crops or pasture from the land. In less than a decade, the soil is usually exhausted and left to wash away with the torrential rain. In North America, clear-cut logging denudes mountain slopes, leaving exposed soil to wash downhill, clogging creeks and rivers and choking spawning beds. Each creek and drainage has its own genetically distinct population of salmon, and once they are wiped out they are gone forever. On the prairies, modern farming methods sometimes wipe out willow bluffs and wetlands, leaving the earth open to wind and water erosion. Most of this eroded topsoil is not coming back – it is borne away by rivers and swept out to sea.

The hopeful news is that some of this soil erosion can be reduced easily and inexpensively just by changing the way we manage the land. The expression "cheap as dirt," says a lot about our wrong-headed ideas about soil. Dirt is not cheap at all. Without it, most animals could not exist. In recent books like *Collapse*, scientists point out that healthy ecosystems are the very foundation of human society. In its own way, Fort Whyte is trying to change common misconceptions about soil.

Fort Whyte's Prairie Soil Diorama was created by the nationally acclaimed dioramist Betsy Thorsteinson. It begins with a five foot deep cross-section of native prairie, showing the three essential layers of earth, the deeply-rooted grasses and forbs, and the rodents, worms and insects that mix and aerate the soil. With exquisite detail, the three-dimensional dioramas depict the changes in one piece of prairie landscape from pre-European times to the present day. Thorsteinson created each tiny animal and plant by hand, and the landscapes have a fanciful quality that helps us imagine what Fort Whyte itself might have looked like hundreds of years ago. Dioramas show how mechanized farming has permanently altered the prairie ecosystem and left it vulnerable to erosion. The final diorama shows how new farming techniques are conserving the topsoil with shelterbelts, healthy wetlands, contour ploughing and crop rotation.

"Manitoba might be an important agricultural province," says Elliott. "But most of its political clout and most of its citizens live in the city. The city is also our main client base, so we've got an obligation to tackle these subjects."

In keeping with its early incarnation as a wildlife refuge that both seeks to educate and entertain, Fort Whyte offers diversions like The Aquarium of the Prairies and the Richardson Rrrrun toboggan slide as entertainments, and exhibits like the soil diorama to deliver a more serious message. "From the start, our challenge has been to lure people away from their television sets," says Elliott. "Once they're here, we hope to raise their awareness of important environmental causes like soil conservation."

POST CARD FROM THE PAST

It's the end of winter here at Fort Whyte and the grounds are dotted with the little snow houses we call quinzhees.

The school kids love building them. But not many of us get a chance to actually sleep in one, so last week I decided to try it. The quinzhee was ten feet wide and six feet high, large enough to accommodate three adults.

Before settling in for the night we covered the exposed grass with plastic sheets, then a layer of foam sleeping pads, then some sheepskins to insulate us from the frozen ground. We slept in two sleeping bags with wool blankets in between, and took great care to ensure that no snow fell into the sleeping bags when we opened them! Toques and heavy socks completed our ensemble, and only had to shift position once in a while when our noses got cold. It went down to minus 23 during the night, but we were quite comfortable. In the morning the wind was howling outside, and after sticking my head outside, I was tempted to climb back into the quinzhee. All in all, I recommend it. It's quite a treat to experience the snug, silent world where the voles and other mammals live during the winter.

– Wendy Loly, Director of Education 1987-95

Wendy Loly (L), leading a spring Volunteer Interpreter training workshop

Consider...

Try to select a single story that captures the essence of Fort Whyte. You may as well ask me to select one star in the Milky Way. I can only express my impression of this remarkable place, its programs, the people, by painting you a picture with words of some of the images on this broad canvas. Each day brings a new composition. I have had the good fortune to have played a role in some, but Nature ultimately is the master of paint brush.

A winter sky so clear and blue with puffs of smoke rising from the chimney at the Sod House. Volunteers are boiling up syrup on the wood stove to make snow taffy for a Winter Fun Day.

A warm June morning and a woman arrives with a box in her arms to the Visitor Reception Desk. More orphaned mallard ducklings delivered to the safety of the waterfowl room.

Two dozen colourful bicycles claim every railing and post on the outdoor plaza of the Alloway Reception Centre. The Grey Hares bicycle club has ventured out on Fort Whyte forest trails for a morning pedal and nourishment from the Buffalo Stone Café.

A tiny pair of unclaimed shorts lies on the floor in the public washroom. Day Campers are having too much fun to worry about belongings.

Heads are down and backs are bent in the Biodiversity Garden. The Friday morning gardening volunteers are trimming hedges, pulling weeds and labeling native flowers.

School children stand as if to salute with their pointer and index fingers raised. 'These are your two snake petting fingers', the Interpreter explains, holding the wriggling garter.

Coffee cups clink on the tray as they are carried into the library. Six members of the 'mail crew' are busy at work sharing stories and laughs as they fold and label member newsletters.

'Martin 16' sailboats dot Lake Devonian while wheelchairs sit empty on the dock. Disabled sailors are catching the freedom of the wind.

Volunteers smile as they pass around the table a large brown envelope. Precious school children's fan mail complete with crayon impressions and scrawled script thanks them for their field trip.

Families hold the hands of toddlers at they travel down the teetering floating boardwalks. An unconcerned painted turtle slips below the green duckweed surface of the marsh.

Steam rises from the frosty nostrils of Clydesdale horses. Sleigh rides in the forest make for memorable winter Sunday fun.

An orange sky with black silhouettes brings hundreds of visitors out to Fort Whyte. Sunset migration evenings are the best show in town.

– Maureen Krauss

Thousands of visitors flock to Fort Whyte for sunset every fall to witness the natural spectacle of fall migration.

CHAPTER 11

REACHING NEW HORIZONS

The winter of 1987 was a good one for Fort Whyte. The new programs launched several years before were starting to make a big difference, and the Centre was hosting more than 50,000 visitors a year, well ahead of its projections.

Of those, 20,000 were students. In the words of a very pleased Alan Scarth, "On one special Sunday more than twelve hundred people showed up for ice fishing, skiing, skating, and just enjoying the bright sun and companionship. The parking lots were jammed, our entrance road was covered with cars along both sides of its length, and the lakes were covered with cheerful enthusiasts."

The year 1987 was also a crucial date in the global environmental movement. The United Nations had formally recognized that the planet's natural systems were in serious trouble, and it launched a massive and far-reaching enquiry called "The World Commission on Environment and Development." A medical doctor and former Prime Minister of Norway named Madam Gro Harlem Brundtland was appointed to chair the Commission, and she became such a strong and impressive chairperson that the inquiry gradually adopted her name, and its final publication became simply known as "The Brundtland Report."

Published in 1987, it marked a decisive moment in the history of the world. There had never been such widespread agreement in the scientific community that the planet was in big trouble. After the release and widespread discussion of the Brundtland Report, it began dawning on governments around the world that these problems needed to be fixed very soon or the future of all wild species and indeed humanity itself was in jeapordy.

Not everyone was happy with the Brundtland Report. Some pronounced it too aggressive in its recommendations, many others thought it not aggressive enough. Springing as it did from a coalition of governments, it placed all the responsibility for change in the laps of government, ignoring the work of small, privately funded organizations like the Fort Whyte Centre, the Nature Conservancy, and Ducks Unlimited. But the Brundtland Commission made one crucial and long-lasting contribution to the history of mankind. It popularized the concept of "sustainable development."

Sustainable development is a popular term nowadays. Only in the late 1980s and 1990s did it become a widespread concept, and as it became popular it brought conservation into the forefront of world affairs. In brief, sustainable development can be defined as "balancing human needs with the protection of the environment, so that these needs can be sustained not only in the present but in the indefinite future."

Sustainable principles can be used in various fields of development – the environment, the economy, and society at large – but the principles are similar in each constituency. The idea is that we should never take more from the environment (or from society, or from the economy) than can be replenished by natural growth. If we are drilling water wells, for example, sustainable development dictates that we mustn't take more than underground water supplies can replace. If we are drawing money from a personal or national investment fund we may take the interest but not the capital. The same principles hold true for devising a foreign policy, enforcing salmon quotas, or helping to rebuild a society in a war-ravaged foreign land. By the rule of sustainable development, it all comes down to the same practical question: Can we justify this in the long term?

Fort Whyte has been in the business of sustainable development for a long time. (It was in fact a realization that Manitobans could not continue their aggressive consumption of the province's natural resources that prompted Alan Scarth, Don Muir, and the other founders of the Wildlife Foundation of Manitoba to sound the alarm back in the 1960s.) But as the new millennium loomed ahead, Fort Whyte, like many other organizations, was thrown into a state of self-evaluation. How could it help to make sustainable development not just a concept but a working principle in the lives of Manitobans?

Incredible progress had been made in the fields of technology, law, medicine, and other constituencies. But many reasonable scientists, futurists, and philosophers feared that humanity was driving the world towards a massive environmental collapse, which might result in the extinction of many species including our own. Some radical thinkers even made the argument that extinction of *Homo sapiens* might not such a bad idea.

Faced with these sobering possibilities, senior management and the Board and the other team members at Fort Whyte took a good look at where the Centre was and where it needed to go. "We clearly needed to expand on a number of levels," says Elliott. "As far as the actual size of the property, we were getting so many visitors that it was obvious to us that Fort Whyte was being loved to death. The deer trails were turning into people trails. We were becoming more successful, but ironically that was creating its own problems, in that our small chunk of bush was being pounded by 100,000 visitors per year. We were like a tiny microcosm of the larger world, with many of the same problems. By the late 1990s it was obvious that if Fort Whyte was going to maintain its ecological integrity and survive as a business we needed to expand our habitat base and facilities."

On April 22nd, 1999, Earth Day, Fort Whyte unveiled its new expansion vision entitled "Reaching New Horizons." The main goals of the campaign included 440 new acres of forest, field and lake, a new reception centre, a café, a bison prairie enclosure, an tipi encampment, and the development of a trail network connecting to the Assiniboine Forest. With the promise of a major millennium grant from The Winnipeg Foundation, Fort Whyte invited Manitobans to help finance the expansion, and they did. "I have to say that raising money for 'bricks and mortar' in Manitoba has never been a major problem," says Elliott. "But you must have a vision that people can identify with, and it must be a sustainable one. Winnipeg is not a big city, but people who live here are very passionate about making it a good community for raising and educating kids. A lot of Winnipeggers feel a real sense of ownership about Fort Whyte, which is entirely appropriate, since they helped build it."

Ms. Brundtland is still involved in global development. **With world leaders like Nelson Mandela, Desmond Tutu, Kofi Annan, Jimmy Carter, and Li Zhaoxing, she belongs to an exclusive team called "The Elders." Formed in 2007, and funded by donors like Richard Branson and Peter Gabriel, The Elders roam the world with a vow "to support courage where there is fear, foster agreement where there is conflict, and inspire hope where there is despair."**

Carl Nelson's seminal sketch of the future visitor reception centre

Bill Elliott on Lake Devonian with the Alloway Reception Centre under construction, 2000

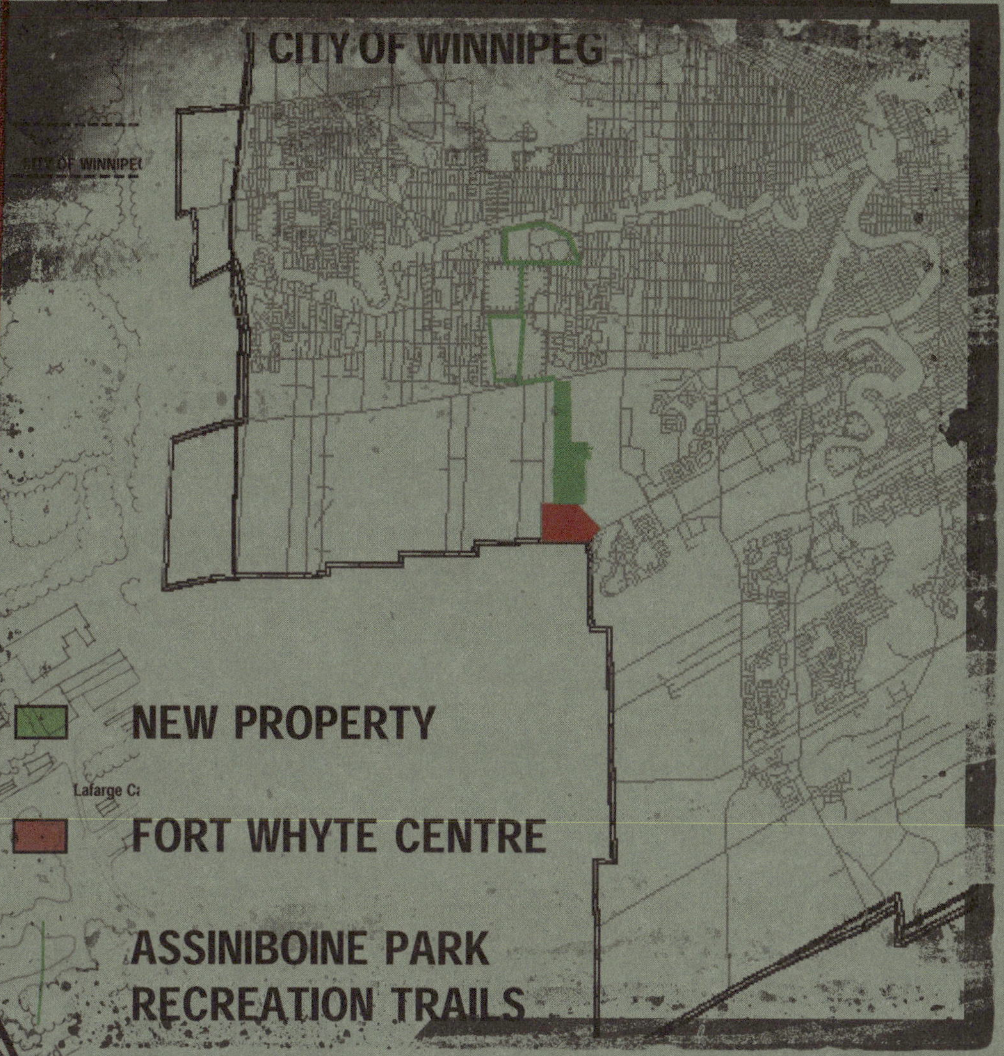

As the millennium approached, plans were completed for the major component of the Reaching New Horizons program – a new reception building. During the winter, construction began, and by springtime new walls and roof joists were rising from the prairie. The new Alloway Reception Centre officially opened its doors in September 2000, and the public, had a chance to see how Carl Nelson, Tom Monteyne, and Dean Syverson had translated the message of conservation and sustainable living into a building that was bright, inventive, spacious, practical, and in harmony with the landscape.

The Alloway Reception Centre is a dramatic contrast to the original Interpretive Centre. The earlier building has a sloping cedar roof facing north, and large, heat-trapping windows facing south. The new building has many different materials, all working in harmony through complex wall and roof planes. Carl Nelson once again supervised the architectural design work, and as he described it: "The building faces the cold north winds with a monolithic wall of Manitoba Tyndal limestone, detailed to resemble dry-laid stone walls and to appear as a geological anomaly on the prairie. This is reinforced by the overall organizational strategy--the delineation of a series of planes, volumes and elements in a dynamic composition. Its apparent randomness intends to suggest that the building form has resulted from tectonic activity at the site, with dynamic soil pressures forcing piles upward out of the ground, causing the roof planes to shift and walls to tilt, suggesting that the building has been bolted together, patched up, and stabilized."

As you walk into the building the first thing you notice are the beefy support posts and ceiling rafters, which are made from parallel strand lumber (PSL), a type of structural wood made by gluing together strands of fibre. This fabrication method allows lumber producers to utilize smaller trees and the entire tree instead of selected, knot-free lengths of old growth timber. The roof deck is composed of wood panels constructed by a similar process, only with fibres of aspen, which are a quick-growing tree not normally valuable for lumber. The windows and doors are made from old cedar hydro poles. These poles, although somewhat labour intensive to recycle, produce beautiful dimensional lumber which doesn't require stain, oils, or other preservatives. Left exposed to the weather, it settles into a natural-looking grey finish that is virtually impervious to rot.

The building is heated and cooled by a geothermal system of coils resting on the bottom of Lake Devonian. This quiet, energy efficient method of exchanging heat in winter (and summer) is aided by large, south facing windows that passively gather solar heat. Water to maintain the gardens and trees is collected from the roof during rainstorms, which in many cases reduces the need for electrical power to run pumps. In any case, most of the plants and trees are hardy native prairie species that require little or no babysitting. The front reception desk was milled by Palliser Furniture from mature aspens – lumber hauled from Fort Whyte's own property. The flooring was constructed of Marmoleum – an environmentally smart product made from cork and Manitoba flax seed. Billed as "an exhibit unto itself" the Alloway Reception Centre not only provides a venue for environmental education but serves as a thought-provoking example of state-of-the-art environmental design.

Manitou Buffalo Rubbing Stone, 1905. R.N. Lea, on his Pembina Valley homestead during an afternoon prairie chicken hunt, pauses by this magnificent, authentic rubbing stone now resting on the south plaza of the Alloway Reception Centre.

THE BUFFALO RUBBING STONE

In 1878, as the last remnants of the great bison herds were fading from the Great Plains, an English dairy farmer named Richard Lea arrived in Manitoba, hoping to find a homestead where he and his family could start afresh. Travelling west on the old Commission Trail, he found an ideal farm site in southwestern Manitoba, with a freshwater creek and abundant grazing land. Leaving instructions with a local man to build him a house, he went back to England and fetched his family. They returned two years later and started farming. The nearest community was a village called Manitou.

"It must have been beautiful country then," says his grandson Percy. "It's still very beautiful, with rolling hills and lots of wildlife, although I guess by then the buffalo had all been chased off. My grandfather's house had one of the biggest fireplaces you've ever seen, and the house is still standing, although no one lives there. They built a church that's still standing too, and people still get married there. There's a lovely cemetery out back and the whole Lea family is buried there.

When plans were underway to build the Alloway Reception Centre, Bill Elliott put the word out in rural Manitoba that Fort Whyte was looking for an authentic buffalo rubbing stone. An agriculture representative did some research and one day the telephone rang in the home of Percy, age 73. "Bill Elliott called me and said that he'd heard that we had a buffalo stone on the farm. I told him yes we had a beautiful big one. He said he wondered if we would consider donating it to Fort Whyte."

It's estimated that 30-60 million bison once roamed the Great Plains. In spring, they would shed their heavy coats of winter fur by rubbing against any large, rough object they could find. Rocky outcrops are few and far between on the prairies. But here and there are large boulders, or "erratics," that were deposited by the vast continental ice sheet 10,000 years ago. Bison loved these boulders, and used them so much that the sharp edges became worn and polished. "We actually have a few of them on our land," says Percy Lea. "You can tell a buffalo stone because it's worn glistening smooth and it has a ditch around it from the countless buffalo walking around it."

Lieutenant Governor, Peter Liba, dons an authentic Winnipeg Police Department issue buffalo coat at the announcement of Fort Whyte's new visitor reception centre.

Percy Lea talked to his brothers about donating the stone to Fort Whyte. "They had mixed feelings," he says. "They felt it was part of our family heritage. But I said, 'Now listen boys, that's true. But we're not getting any younger. What if someone buys this further on down the line and they just bulldoze that stone into a creek. Then all that history will be gone.'"

The rubbing stone is a symbol of Manitoba's ancient heritage. It connects us to the pre-European past, when millions of bison roamed the plains, and it connects us to the native people who once lived a life closely interwoven with the lives of the bison. And the stone is also a symbolic welcome to the bison that now make their home at the Centre.

In a tribute to those historical connections, native elder Lawrence Houle blessed the stone in a traditional sweetgrass ceremony. Lieutenant Governor Peter Liba was on hand too, wearing an immense buffalo coat loaned from the Winnipeg Police Museum. Touching the polished edge of the stone, Liba affirmed the commitment of Fort Whyte and its supporters to safeguard the natural environment for future generations.

Another big addition to Fort Whyte was the bison herd. When the Reaching New Horizons campaign was launched on Earth Day, 1999, the acquisition of a bison herd was also announced, and the public began getting very excited about the return of these legendary beasts to the environs of Winnipeg. What better symbol of Fort Whyte's commitment to the future than this majestic animal from centuries past?

It's one thing to announce the arrival of a bison herd, and quite another to go out and find the animals. In 1998, Bill Elliott got together with site manager Ken Cudmore and tried to figure out how they could get a herd. The Fort Whyte staff members were going on a staff retreat to the Elkhorn Ranch so Elliott and Cudmore used the opportunity to talk to the Riding Mountain National Park officials about their bison herd and take a tour of the facility. Says Elliott, "The Riding Mountain animals are treated like a wild herd. The park officials try to duplicate the natural way that the bison would live if there was no park there. If a wolf kills a young bison, the officials don't get involved. It's very hands-off, and it's quite different from what we had in mind. But we learned a lot about the interpretive angle, how they allow people to observe the bison without detracting from the natural experience."

K 5052 TMX 2

BISON

When they got back to Winnipeg, Bill Elliott called Bo Wohlers, a long time friend and supporter and owner of the Rockwood Bison Ranch. He offered to come by the Centre and show them what they would need for fencing, corrals, squeeze chutes, pasture and so on. "We learned that you can't fool around when it comes to fencing," says Elliott. "They're extremely powerful and they like to go wherever they please. So we sent away to Alberta for extra strength page wire and built a super-sturdy eight foot high fence. Then we seeded the pasture with the right kind of forage, let it grow, and began examining the thorny issue of how we were going to find the money to buy the bison."

Bison were in high demand at the time and a pregnant heifer sold for up to $6,000. The Centre didn't have that kind of money so Elliott tried to get some for a Fort Whyte discount, i.e., for nothing. He sent letters to the members of the Manitoba Bison Producers Association soliciting donations of surplus animals, as if there was such a thing. In the meantime, Bo Wohlers stepped up and offered to loan them a small herd of yearlings.

"It was quite something when they arrived," says Ken Cudmore. "A truck came trundling down the road with a couple of dozen yearlings onboard. It was a big event. The media were there with cameras and were all hoping to get some great photographs of the young bison galloping through their new enclosure. But we'd done such a good job planting the pasture forage that it was almost waist high. The young bison piled off the truck and ran into the grass and basically disappeared."

Bo Wohlers reclaimed his bison calves after they matured, but then several members of the Manitoba Bison Producers Association stepped up to the plate and donated five animals. Around the same time, the park wardens at Riding Mountain decided their herd was getting overpopulated so they offered to give Fort Whyte 17 adult bison. Soon the herd stabilized at about 24 animals, which is where it remains today. Ordinarily the herd consists of two bulls and a dozen breeding cows, each of which normally "drops a calf" in May or June.

Every spring, there is great anticipation as the cows prepare to give birth to a new crop of bison calves. Visitors flock to the viewing mound or the Saturn Family Treehouse to search for the awkward, cinnamon-coloured newcomers. The bison tend to stay well away from the people and vice versa. Bison are potentially dangerous and have to be viewed from a distance, especially during calving time. The bison cow has a lot invested in her calf and she will not tolerate intruders. The gestation period for bison is about the same as that for humans (285 days). Before she gives birth, the mother bison scans the prairie to make sure the coast is clear before she goes into labour. Bison calves are vulnerable to wolves and other predators for the first few hours after they are born, and the cow can delay labour if she's not sure the situation is safe.

As soon as the calf is born she licks its coat clean and encourages it to nurse. A new bison calf weighs about 50 pounds, and after a few hours of nursing it can gallop around and keep up with the herd if required. One of Ken Cudmore's many responsibilities is to take care of these bison calves. On his four-wheel drive bike he goes in amongst the herd and checks on their food supply, their water and the general health of the new calves. Things usually go smoothly but not always. Once, a cow gave birth to a stillborn calf and it jammed on the way out. He and Bill Elliott had to herd the distressed cow into a squeeze chute and work for several hours, trying to deliver the dead calf. Finally they succeeded, and the mother escaped with her life. "But it was gruesome work," says Bill Elliott.

"Bison are very peaceful animals," says Ken Cudmore. "Until you try to get them to do something they don't want to do. Then their eyes cock back and they start getting ornery."

Cudmore says that when he gets off the bike he keeps an eye on the bulls, which can weigh as much as a pickup truck and tend to be hostile to interlopers. "They know me, so they're pretty calm when I'm in amongst them. But still, you can't take liberties. I don't go too far from the bike and I don't ever turn my back on them."

Has he ever been injured by a bison? Well yes, in fact; a bull once put him in the hospital. It was a quiet spring without any births in the herd, and whenever he checked there wasn't much to report. The year before, the first bison calf had been born in late April – on Earth Day. So when that same time arrived the following year he watched the herd with daily anticipation, "ready to hand out the cigars," as he puts it. But nothing happened. The end of May came and went, but there was still no sign of cinnamon-coloured calves. "I was beginning to question the masculinity of our two breeding bulls," he says. "But not to their faces."

At the end of June a calf was finally born, and that blessed event seemed to kick off a baby boom that resulted in seven more calves hitting the ground over the next three weeks. One afternoon he was in amongst the herd in his truck. And as he was leaving, opening the gate, he noticed movement in the high grass along the fence. "I thought maybe it was a jackrabbit or something," he says. "But it was this tiny little bison calf feebly stumbling along the fence line."

Bison usually stick together, but this calf was 400 yards from the rest of the herd. He loaded the helpless little newborn symbol of Manitoba into his truck and went to find the negligent mother. "It was obvious which one she was, but even though I tried to interest her in the calf she totally ignored it. She was your classic dysfunctional mother. I don't know if she was too young and inexperienced or what her problem was, but it was pretty clear that she wouldn't be mothering this calf anytime soon."

Field Studies

What to do now? He had been up against this kind of dilemma before, and although he was reluctant, he already knew the rest of the script. A short while later, he opened the back door of his house and called to his wife, delivering the dreaded words she'd heard so many times before, "Honey, I brought a little surprise home."

Cudmore had raised many deer fawns before so he knew the procedure. He filled several baby bottles with a 50-50 mix of evaporated milk and water and began feeding the calf. But he soon discovered that there is a huge difference in appetite between a five pound whitetail fawn and a 50 pound bison calf. A tiny fawn can barely handle half a bottle of milk mix. The bison calf sucked back six.

Within three days the little bull calf was strong and full of energy, romping around the backyard with Cudmore's black Labrador. His wife had recently planted a nice spring garden but Cudmore says that after a few days "it looked like it had been dug up with a roto-tiller." It was clear that the yard would be too small for the rambunctious little beast. So after the initial crisis was past, Cudmore found the calf a temporary home with a local farmer. When the calf was large enough and strong enough, Cudmore picked him up with his truck and took him back to the pasture at Fort Whyte. "It's time for you to start being a bison again," he told the young animal.

He leaned over to say goodbye and scratch the calf behind the ears. Bison calves stimulate the mother's milk to drop by butting the udder with their foreheads, and maybe at this point the little bull thought Cudmore was his long-lost mommy. "Sometimes they butt the mother's udder so hard they almost lift her hindquarters off the ground," says Cudmore. As he leaned over and whispered his goodbyes, the calf delivered a head-butt.

When Cudmore regained consciousness, he was lying on his back on the ground with blood pouring from his mouth. The calf had hit him so hard his teeth had punched right through his lip.

He went to the hospital, where the admitting nurse expressed dismay at the damage to his mouth. "My god, what happened to you?"

"I told her I'd been attacked by a bull bison," says Cudmore. "I didn't tell her how small it was."

AWARDS IN EDUCATION

Bob Drummond, Vicki Daubaras and Terry Brownlee accept the Mayor's Volunteer Services award on behalf of Fort Whyte's 300+ volunteers

In the autumn of 2005 Fort Whyte received two awards. The first was the City of Winnipeg Accessibility Award, commending Fort Whyte for its ongoing efforts to provide "community leadership in the application of universal design and accessible outdoor landscape." The Centre was nominated for this award by a group called Disabled Sail Manitoba, which encourages and assists disabled Manitobans in learning to manage a sailboat. Disabled people often require special fittings and controls on their sailboats, and Disabled Sail Manitoba has been a pioneer in this regard, with uniquely outfitted boats and regular sailing classes on Lake Devonian. There are three different programs at Fort Whyte, and they are available to all Manitobans regardless of ability, age or economic means. One parent described the benefits: "The fact that our son can sail on his own makes him feel the same as everyone else. He is so proud."

The second award related to Fort Whyte's work in education. Every year, Fort Whyte hosts approximately 30,000 students from grades K-12, engaging them in outdoor programs that relate directly to their school science or social studies curricula. It's one thing to learn a principle in the classroom, but quite another to see it in real life. As one student put it, as she dip-netted tiny aquatic monsters from the marsh and released them into a plastic container of water, "I love catching them and identifying them on the chart. When you're just looking at the pictures in a text book it's kind of boring."

The science teachers commended the staff and volunteers at Fort Whyte for their willingness to work "outside the box," in designing programs that are tailored for the many different groups that show up in the parking lot every day. Fort Whyte has become a sort of open-air laboratory for students, with timely themes (like climate change) that relate directly to issues in the daily news. It's not all scientific learning and serious research either. Students can study outdoor skills, animal behaviour and wilderness survival. High school kids, for example, can enroll in environmental programs that involve two days in the woods at Fort Whyte, with a sleepover at the cabins at the Siobhan Richardson Field Station. The Science Teachers' Association of Manitoba celebrated these and other programs by giving Fort Whyte the 2005 Science Achievement Award. It was a much appreciated gold star to finish off the year.

On the Farm

FortWhyte
FARMS

Growing Youth, Food & Community

Fort Whyte believes that sustainable communities are built on three essential and interdependent cornerstones: a healthy natural environment; a healthy, vibrant economic environment, and a healthy and just social environment. With 640 acres of urban natural resources providing countless educational, recreational, and social enterprise opportunities, Fort Whyte is uniquely positioned to engage all members of the community in innovative programming that builds and nurtures sustainable communities.

As a catalyst in the creation of sustainable communities, the Fort Whyte Farms Program engages marginalized, inner-city youth in innovative urban agriculture projects. Products such as honey, vegetables, fruit, pastured poultry, bison meat fruit and other locally produced fare – are grown, harvested, and sold at market.

Since 2004, Fort Whyte has been working in partnership with a network of respected inner-city schools and youth serving agencies on a weekly basis with groups of youth ages 14-18. Fort Whyte Farms offers young people increased self-reliance while providing insightful experience in business operations. Through hands-on training and experience in agricultural projects, greenhouse management, animal husbandry, retail and marketing, farm interns are equipped with genuine, meaningful employment skills such as team work, leadership, initiative and discipline. Youth leave the program with a sense of hope, place, and purpose, combined with valuable experience in sustainable urban agriculture.

CHAPTER 12

THE EVOLUTION CONTINUES

The famous Harvard paleontologist Andrew Knoll says "When you have the right conditions, life emerges fairly rapidly." If scientists and the rest of us can learn one thing about Nature, it's that very, very little is required of us to ensure a healthy natural environment.

Nature is not delicate, or temperamental. All species, from a tiny whirling rotifer to a one-ton bull bison come equipped with a hard-wired passion to survive, breed and multiply. They don't need our help or our programs or our interpretation in order to survive. They'll do just fine or even better without us. Many species, it's true, are in danger, and tens of thousands have become extinct. Species that fail to adapt, or that somehow outgrow the carrying capacity of their habitat just disappear. But they don't go willingly. Starvation, conflict, and epidemic disease are Nature's remedies for evolutionary designs that don't work out.

Currently, the human population of the world is increasing by over 200,000 per day. By 2050, the world will carry more than 9 billion people, many of whom will attempt to inhabit landscapes drastically degraded by development and global heating. A hundred years ago, not many people believed that an apocalypse was coming. Two centuries ago, the world had its problems, but there was no widespread scientific consensus that our species was headed for a reckoning. Today, that fear is widespread. Even schoolchildren sense it. We have a few generations to change the way we live. And if we can't dramatically reduce our impact on the planet, Nature will do the job for us, by dramatically reducing the human species.

In the early days of the Winnipeg Wood Duck Program, Scarth and Muir and the other patriarchs of the fledgling Fort Whyte Centre were trying to share a simple concept with the citizens of Winnipeg – if you provide the basic conditions for life, wild creatures will do the rest. As the ghost of Shoeless Joe Jackson once whispered in the farmer's ear, "If you build it, they will come." But that's not entirely the truth, and the founders knew it, and still know it. We don't have to build habitat to make the world okay. We

"Connecting Communities Through Nature" FortWhyte's portion of the Trans Canada Trail forms the spine of its 7 kilometre network of nature trails

only have to stop degrading it. The burning of fossil fuels and the resulting global heating are now at the forefront of our concerns.

The story of Fort Whyte is therefore not so much a story of construction as one of deconstruction. Lake Devonian, the Bison prairie, the Green Corridor; these and other Fort Whyte projects are in fact experiments in unbuilding. They are rents in the fabric that allow Nature to pour back in. Once animals like bison and deer and Canada geese get established they need no more help or maintenance. Thanks to them, Winnipeg's southwest corner has become an oasis of biodiversity. And given the crisis that the World is in, that's a good thing.

When the Green Corridor is completed, a mixed swath of tall grass native prairie, aspen forest, lake, and prairie marsh will stretch from Assiniboine Park all the way down through FortWhyte Alive to LaBarriere Park – over 20 kilometres of green space for wildlife and future generations. In the autumn of 2007, Bill Elliott took part in a cycling rally to celebrate the opening of FortWhyte's flagship section of that long green trail. One warm sunny Sunday morning in late September, with the leaves turning yellow and hundreds of geese yelping overhead, Elliott and 150 other cyclists marshaled at a prearranged meeting point in the Assiniboine Forest and set out to cycle the route.

Active Living

Maureen Krauss was waiting at FortWhyte to greet the cyclists with a hundred tiny pails filled with crushed limestone so that each could symbolically contribute to the completion of the path. "The opening of the Alloway Reception Centre was an important milestone for us," says Elliott. "But it was really just a part of the larger dream of connecting all these different spaces together into the Green Corridor. So as we cruised down that trail it was a big moment for me and FortWhyte."

Elliott was wearing a helmet and the other cyclists assumed he was just another outdoor enthusiast enjoying this gorgeous autumn Sunday. When they topped the hill at the north end of the property Elliott squeezed his brakes and pulled to a stop, held back for a moment so he could watch the long-awaited and inspiring spectacle of the 150 cyclists flowing down the hill and into the tall grass prairie with the bison herd. "It was like someone had spoken to the bison," he says, "because the entire herd had turned out to watch us."

With the grazing bison, and the distant tipis and sod hut, he had a strong sense that he was looking at a landscape that was gradually growing into the past. Later on that afternoon, over 1,000 people showed up to celebrate the FortWhyte of today and tomorrow. And that evening at dusk, ten thousand geese settled on the lakes, drawing the curtain on another day in the incredible life of FortWhyte Alive.

Bill Elliott, President & CEO

640 acres of urban greenspace, preserved for all...forever.

ACKNOWLEDGEMENTS

While FortWhyte Alive has afforded generations the opportunity to experience and enjoy the natural environment, generations of the Cyprik Family have reciprocated with genuine interest and generosity in making this book possible. Deepest appreciation is extended to Bob and Gail Cipryk, daughter Nancy, her husband Greg Klassen, and their young daughters, Lura and Natalie representing the generation to whom we all pass the environmental torch.

Posthumous thanks to a longtime friend, Jack MacIver, for his combination FortWhyte wildlife photo album — natural cookbook idea that eventually evolved into this more ambitious undertaking.

And finally, thanks to Alan Scarth, FortWhyte's founder, first Chairman of the Board and true environmental pioneer, for his essential and vivid recollections of the critical and colourful formative years.

The Author

Over twenty-five years Jake MacDonald has produced ten books and hundreds of stories for various magazines and newspapers. Many of his books and articles have been recognized with national writing awards. He divides his time between Winnipeg, Toronto and Minaki, Ontario.

Design & Typography

Relish DESIGN STUDIO LTD.

Illustrations

Linda Fairfield
David Morrow

Photography

Christian Artuso
Ian Barnett
Joe Bryska
Ken Cudmore
Mike Deal
Michelle Eldridge
Bill Elliott Jr.
Gary Frankel
Jennifer Franzin
Marc Gallant
W.C. Gomm
Mike Grandmaison
Garry Hilderman
Phil Hossack
Roland Huemmer
Henry Kalen
Ted McLachlan
Jolene Mulaire
Jim Penner
John Pura
Mark Rash
Lisa Rumble
George Siamandas
Helen Scarth
Andrew Sikorsky
Natalie Smith
Tony Smith
Stan Swanson
Larry Trupp
Len Van Roon
Isabel Wolinsky
The Muir Family
World Ark
Winnipeg Free Press
Archives Manitoba
Glenbow Collection

... and the many other FortWhyte friends who have enjoyed nature from behind the camera lens.